INTRODUCTION TO
Concurrency in Programming Languages

Chapman & Hall/CRC
Computational Science Series

SERIES EDITOR

Horst Simon
Associate Laboratory Director, Computing Sciences
Lawrence Berkeley National Laboratory
Berkeley, California, U.S.A.

AIMS AND SCOPE

This series aims to capture new developments and applications in the field of computational science through the publication of a broad range of textbooks, reference works, and handbooks. Books in this series will provide introductory as well as advanced material on mathematical, statistical, and computational methods and techniques, and will present researchers with the latest theories and experimentation. The scope of the series includes, but is not limited to, titles in the areas of scientific computing, parallel and distributed computing, high performance computing, grid computing, cluster computing, heterogeneous computing, quantum computing, and their applications in scientific disciplines such as astrophysics, aeronautics, biology, chemistry, climate modeling, combustion, cosmology, earthquake prediction, imaging, materials, neuroscience, oil exploration, and weather forecasting.

PUBLISHED TITLES

PETASCALE COMPUTING: Algorithms and Applications
Edited by David A. Bader

PROCESS ALGEBRA FOR PARALLEL AND DISTRIBUTED PROCESSING
Edited by Michael Alexander and William Gardner

GRID COMPUTING: TECHNIQUES AND APPLICATIONS
Barry Wilkinson

INTRODUCTION TO CONCURRENCY IN PROGRAMMING LANGUAGES
Matthew J. Sottile, Timothy G. Mattson, and Craig E Rasmussen

INTRODUCTION TO
Concurrency in Programming Languages

MATTHEW J. SOTTILE
TIMOTHY G. MATTSON
CRAIG E RASMUSSEN

CRC Press
Taylor & Francis Group
Boca Raton London New York

CRC Press is an imprint of the
Taylor & Francis Group an **informa** business

A CHAPMAN & HALL BOOK

MATLAB® is a registered trademark of The MathWorks, Inc. For product information, please contact:

The MathWorks, Inc.
3 Apple Hill Drive
Natick, MA 01760-2098 USA
Tel: 508-647-7000
Fax: 508-647-7001
E-mail: info@mathworks.com
Web: www.mathworks.com <http://www.mathworks.com/>

Chapman & Hall/CRC
Taylor & Francis Group
6000 Broken Sound Parkway NW, Suite 300
Boca Raton, FL 33487-2742

© 2010 by Taylor and Francis Group, LLC
Chapman & Hall/CRC is an imprint of Taylor & Francis Group, an Informa business

No claim to original U.S. Government works

Printed in the United States of America on acid-free paper
10 9 8 7 6 5 4 3 2 1

International Standard Book Number: 978-1-4200-7213-6 (Hardback)

Library of Congress Cataloging-in-Publication Data

Sottile, Matthew J.
    Introduction to concurrency in programming languages / Matthew J. Sottile, Timothy G. Mattson, Craig E Rasmussen.
        p. cm. -- (Computational science series)
    Includes bibliographical references and index.
    ISBN 978-1-4200-7213-6 (hardcover : alk. paper)
    1. Programming languages (Electronic computers) 2. Parallel processing (Electronic computers) I. Mattson, Timothy G., 1958- II. Rasmussen, Craig E III. Title. IV. Series.

QA76.7.S62 2010
005.2'75--dc22   2009031462

Visit the Taylor & Francis Web site at
http://www.taylorandfrancis.com

and the CRC Press Web site at
http://www.crcpress.com

Matthew J. Sottile and Timothy G. Mattson and Craig E Rasmussen

Introduction to Concurrency in Programming Languages

CRC PRESS

Boca Raton   London   New York   Washington, D.C.

Contents

List of Figures

Chapter 1

Introduction

Objectives:

- Introduce the area of concurrent programming and discuss the relevance and timeliness of the topic.
- Investigate some areas where concurrency appears, both in programs and real-world situations.

This book is intended to give an introduction to thinking about concurrency at the level of the programming language. To date, the vast majority of mainstream programming languages are designed to express sequential programs. Current methods for writing concurrent programs in these languages exist primarily as add-on libraries that encapsulate the expression of concurrency in a form that, to the compiler or language, remains sequential. Techniques such as OpenMP provide more information to the compiler in the form of code annotations, but they still suffer from the fact that they are inherently bound to the sequential languages that they are used with. Java is one of the only mainstream languages with concurrency concepts defined as part of the language itself, and even then, most of the facilities for concurrent programming are provided within the standard library and not the language definition. Given that these facilities most often exist outside the scope of language definitions, the acts of optimization, efficient code generation, and correctness checking by compilation and source analysis tools prove to be very difficult.

Language-level constructs created specifically for concurrent programming are necessary to exploit parallel processing capabilities in a general sense that facilitates automatic compilation and analysis. We will discuss a set of these language constructs in this text based on those that have proven successful in the decades of parallel languages research, in the hope that programmers will understand and use them to best take advantage of current and upcoming architectures, especially multicore processors. These will be demonstrated in the context of specific concurrent languages applied to a set of algorithms and programming patterns. Using these constructs requires an understanding of the basic concepts of concurrency and the issues that arise due to them in constructing correct and efficient concurrent programs. We will cover this topic to establish the terminology and conceptual framework necessary to fully understand concurrent programming language topics.

Readers should come away from this book with an understanding of the effect of concurrency on programs written in familiar languages, and the language abstractions that have been invented to truly bring concurrency into the language and to aid analysis and compilation tools in generating efficient, correct programs. Furthermore, the reader should leave with a more complete understanding of what additional complexity concurrency involves regarding program correctness and performance.

Concurrency and parallelism: What's the difference?

Before we dive into the topic, we should first establish a fundamental piece of terminology: what does it mean to be concurrent versus parallel? There is a great deal of press currently about the advent of parallelism on the desktop now that multicore processors are appearing everywhere. Why then does the title of this book use the term *concurrency* instead of *parallelism*? In a concurrent system, more than one program can appear to make progress over some coarse grained unit of time. For example, before multicore processors dominated the world, it was very common to run multitasking operating systems where multiple programs could execute at the same time on a single processor core. We would say that these programs executed *concurrently*, while in reality they executed in sequence on a single processing element. The illusion that they executed at the same time as each other was provided by very fine grained time slicing at the operating system level.

On the other hand, in a multicore system (or, for that matter, any system with more than one processor) multiple programs can actually make progress at the same time without the aid of an operating system to provide time slicing. If we run exactly two processes on a dual-core system, and allocate one core per process, they will both make progress at the same time. They will be considered to be executing in *parallel*. Parallelism is a realization of a concurrent program. Parallelism is simply the special case where not only do multiple processes appear to make progress over a period of time, they *actually* make progress at the same time. We will discuss the details of this and other fundamental concepts further in Chapter 2.

1.1 Motivation

Concurrency is a core area of computer science that has existed for a long time in the areas of high-performance and scientific computing, and in the design of operating systems, databases, distributed systems and user interfaces. It has rapidly become a topic that is relevant to a wider audience of programmers with the advent of parallel architectures in consumer systems. The presence of multicore processors in desktop and laptop computers, powerful programmable graphics processing unit (GPU) co-processors, and specialized hardware such as the Cell processor, has accelerated the topic of concurrency as a key programming issue in all areas of computing. Given the proliferation of computing resources that depend on concurrency to achieve performance, programmers must learn to properly program them in order to fully utilize their potential.

Accompanying this proliferation of hardware support for parallelism, there is a proliferation of language tools for programming these devices. Mainstream manufacturers support existing techniques such as standardized threading interfaces and language annotations like OpenMP. Co-processor manufacturers like Nvidia have introduced the Compute Unified Device Architecture (CUDA) language for GPUs, while libraries have been created by IBM to target the Cell processor found in the Playstation game platform and various server-class machines. The Khronos group, an industry standards organization responsible for OpenGL, recently defined a language for programing heterogeneous platforms composed of CPUs, GPUs and other processors called OpenCL. This language supports data parallel programming models familiar to CUDA programmers, but in addition OpenCL includes task parallel programming models commonly used on CPUs.

In the high-performance computing industry, Cray is developing a language called Chapel, while IBM is working on one called X10 and Sun is working on a language called Fortress. The 2008 standard for Fortran includes primitives specifically intended to support concurrent programming. Java continues to evolve its concurrency primitives as well. The Haskell community is beginning to explore the potential for its language in the concurrent world. How does one navigate this sea of languages, and the tide of new ones that is surely to continue to follow?

1.1.1 Navigating the concurrency sea

Unfortunately, there is only a small amount of literature available for programmers who wish to take advantage of this capability that now exists in everyday devices. Parallel computing has long been primarily the domain of scientific computing, and a large amount of the educational material available to programmers focuses on this corner of the computing world. Computer sci-

ence students often are introduced to the issues that arise when dealing with
concurrency in a typical curriculum through courses on the design of operating systems or distributed systems. Similarly, another area of likely exposure
to issues in concurrency is through databases and other transaction-oriented
server applications where software must be designed to deal with large numbers of concurrent users interacting with the system. Unfortunately, literature
in each of these fails to provide readers with a complete and general introduction to the issues with concurrency, especially with respect to its impact on
designing programs using high-level languages.

Operating systems typically are implemented with lower level languages for
performance reasons and to facilitate fine grained control over the underlying
hardware. Concurrency can be quite difficult to manage at this low level
and must be planned very carefully for both performance and correctness
reasons. On the other extreme are database-specific languages, such as SQL.
In that case, programmers work at a very high level and are concerned with
very specific types of concurrent operations. Concurrency in both of these
cases is very specific to the problem domain, and looking only at these cases
fails to paint a complete picture of the topic. This book aims to provide an
introduction for computer scientists and programmers to high-level methods
for dealing with concurrency in a general context.

The scientific computing community has been dealing with parallelism for
decades, and many of the architectural advances that led to modern concurrent computing systems has grown out of technological advances made to
serve scientific users. The first parallel and vector processing systems were
created to fulfill the needs of scientists during the Cold War of the twentieth
century. At the same time, mainframe manufacturers such as IBM and systems software developers at AT&T Bell Laboratories sought to design systems
that supported concurrent operation to service multiple users sharing single,
large business computing resources. The work in scientific computing focused
on concurrency at the algorithmic level, while designers of multiprocessing
operating systems focused on concurrency issues arising from the interactions
of independent, concurrently executing programs sharing common resources.

The problem most modern readers will find with much of this literature is
that it focuses on programmers using parallelism by working at a very low
level. By low level, we imply that the programmer is given direct and detailed
control over how computations are split across parallel computing elements,
how they synchronize their execution, and how access to shared data is managed to ensure correctness in the presence of concurrent operations. This focus
on low-level programming is, in comparison to traditional sequential programming techniques, equivalent to teaching programmers to use techniques close
to assembly language to write their programs. In sequential programming,
one considers assembly language to be at the "low level" of the hierarchy of
abstraction layers. Assembly language programmers control the operations
that occur in the computer at a very fine grain using relatively simple operations. Operations are specified in terms of individual arithmetic operations,

basic branches that manipulate the program counter to move between relevant regions of executable instructions, and explicit movement of data through the memory hierarchy from the main store to the CPU.

Many decades ago, with the advent of programming languages and compilers, programmers accepted that higher level abstractions were preferable to utilize hardware in a portable, efficient, and productive manner. Higher levels of abstraction associated with a language construct are related to the amount of insulation and protection it provides to the programmer above the underlying hardware. Higher yet are constructs that not only insulate programmers from the machine, but attempt to encapsulate concepts and techniques that are closer to the human description of algorithms and data structures. In addition, higher-level abstractions give the compiler more freedom in emitting code that is specific to a given computer architecture.

Complex arithmetic expressions, such as the computation of a square root or transcendental function, both of which could require detailed sequences of assembly instructions, could be expressed as atomic operations translated to machine code by a compiler. Similarly, the flow of control of a program could be abstracted above rudimentary program counter manipulations to operations that were closer to those used to design algorithms, such as `do-while` or `for`-loops and `if-then-else` conditional operations. Programmers realized, and quickly accepted, that algorithmic tools could assist them in translating algorithms from an abstraction layer where they could comfortably think and design programs to a lower one where the machine could execute equivalent machine code.

The basis of this text is that a similar set of abstractions exists for concurrent programming, and that programmers should start to think at this higher level with respect to concurrency when designing their programs to both achieve portability, high performance, and higher programmer productivity. In a world where concurrency is becoming pervasive, abstractions for concurrency will become increasingly important.

Unfortunately, parallel programming languages that provide this higher level of abstraction to programmers have yet to gain any acceptance outside of niche academic circles. There has not been a parallel equivalent of Fortran or C that was accepted by the broader computing community to build the critical mass of users in order for production-grade compilers and educational literature to be created. Java is the closest that can be found in popular usage, but it has not proven to be generally accepted in contexts where other languages have traditionally been used. Parallel languages inspired by Fortran, C, Lisp, Java, and many others have been proposed, but all have failed to gain widespread usage and acceptance. Parallel languages often have been relegated to the shelves of libraries in high-performance computing literature, and have lived most often in academic circles as research topics with the sole purpose of providing research contexts for Ph.D. dissertations. With parallelism clearly on the desktop, it is vital to change this state of affairs for the benefit of programmers everywhere. Programmers must be provided with tools to manage concurrency in general purpose programs. Language designers have responded in recent years with a variety of exciting concurrent languages.

The key is to focus on the language constructs that provide high level mechanisms for the programmer to utilize concurrency with the same amount of intellectual effort required to design traditional sequential programs. Ideally, building a program that uses concurrency should be only slightly more difficult than building a sequential one. It has become clear that, although many languages for parallel computing have come and gone over the years, there are successful abstractions that persist and appear over and over in languages to reduce the complexity of using concurrency to the developer's advantage. Educating programmers about these successful and persistent abstractions, and teaching them how to apply them to algorithms that are relevant to their work, is key in providing them with the background to be ready to use new languages as they arise. Furthermore, users who choose to remain in traditional sequential languages augmented with add-on tools for concurrent programming who understand these abstractions can implement them in their own programs when true parallel languages and compilers are absent.

1.2   Where does concurrency appear?

Programmers have adapted to concurrency whether or not they appreciate it in many contexts. A working programmer is hard pressed to claim that he or she has not had to deal with it at some point in his or her career. What are the contexts where this may have occurred?

Operating systems

The operating system of a computer is responsible for managing hardware resources and providing abstraction layers that give programmers a consistent and predictable view of the system. Few application programmers are ever concerned in modern times with the details related to how one sends a text file to a printer, or how a block of data traverses a network connection to another computer. The operating system provides interfaces to drivers that abstracts this from the programmer. Similarly, all modern operating systems[1] allow multiple programs to execute concurrently, each accessing these resources in a coordinated way. This has been especially useful with single processor systems where the single processing unit has been multiplexed to provide the illusion of parallel execution in modern multitasking operating systems.

---

[1]Embedded operating systems are the most common modern exception to this.

Concurrency arises in many contexts within operating systems. At a fundamental level, the operating system manages programs that execute at the same time through some form of time-sharing or multitasking. One can take a standard operating system today, run two programs, and they will happily coexist and run independently. The operating system ensures that when these programs are sharing resources (such as I/O devices and memory), they run in a predictable way in the presence of other concurrent processes. By "predictable," we imply that the programs execute in the same way regardless of whether they execute alone on the system or concurrently with other processes. The only significant difference that is likely to occur is an increase in the runtime due to contention for shared resources (such as the processor). From a correctness sense, the system will ensure that programs not sharing and modifying data (such as a file in the file system) will produce the same output regardless of the other processes executing concurrently on the machine.

Correctness is not the only concern in operating system design however. Given a set of processes that are sharing resources, each expects to make some reasonable amount of progress in a given amount of time. This is handled by scheduling algorithms that control the sharing of these resources between concurrently executing processes. In this context, it is important to provide some guarantee of service, or fairness, to all processes. As a result, operating system designers must also be concerned with the performance characteristics of the concurrent parts of their system in addition to their correctness.

Distributed systems

The majority of computer users today take advantage of distributed, network-based services, such as the world wide web or e-mail. Servers on the Web, especially those that are popular, have to deal with the fact that many users may be attempting to access them at any given time. As users expect a high degree of responsiveness, server designers must deal with concurrency to ensure a good user experience on their sites. This requires server programmers to pay attention to details such as:

- How to efficiently service requests as they arrive.

- Preventing connections with varying performance characteristics from interfering with each other.

- Managing data to cache frequently accessed information, and preventing conflicts between clients that make modifications on shared data.

In many instances, network software has performance as a top priority. This performance is defined in terms of responsiveness and performance as seen by clients. Furthermore, most servers manage some form of state that clients gain access to for reading and modification. Correctness of these concurrent operations is critical to maintaining the integrity of this state.

User interfaces

The user interface has long been an area where concurrency has been a core topic. Concurrency is an unavoidable feature of any interface because there are always at least two concurrently operating participants in an interface — the user and the program he or she is interacting with. In most graphical interfaces, the system allows the user to see and interact with two or more programs that are executing simultaneously. Managing the interactions of these programs with the shared graphical hardware and input from the user are areas where the user interface programmer must consider the implications of concurrency.

The user interface system must manage shared resources (such as the display and input device) and ensure that the proper programs receive the appropriate input and that all programs can display their interfaces without interfering with each other. Similarly, programs that are run in a graphical environment must be written to deal with the concurrency that arises in their internal logic and the logic necessary to interact asynchronously with the user.

Databases

Databases commonly are available to more than one client. As such, there are expected instances in which multiple clients are accessing them at the same time. In the case of databases, the system is concerned with providing a consistent and correct view of the data contained within. There is a delicate balance between providing a responsive and well performing system, and one that guarantees consistency in data access. Clever techniques such as transactions can minimize the performance overhead with many synchronization primitives by taking advantage of the relatively rare occurrence of conflicts between concurrently executing client interactions.

We will see that the successful abstraction provided by database transactions has inspired more general purpose methods that can be integrated with programming languages for writing concurrent code. This will be discussed when we cover software transactional memory schemes.

Scientific computing

Scientific computing has long been the primary domain of research into methods for expressing and implementing concurrency with performance as the key metric of success. Unlike other areas, performance was often the *only* metric, allowing metrics such as usability, programmability, or robustness to be pushed to a secondary status. This is because scientific computing often pushes machines to their extreme limits, and the sheer size of scientific problems, especially in their time to execute, makes performance exceptionally important.

An interesting side effect of this intense focus on performance is that many algorithmic techniques were developed in the scientific computing community for common, complex operations that exhibit provably efficient performance. For example, many operations involving large sets of interacting concurrent processing elements that are naively coded to take $O(P)$ time on P processors can actually achieve a preferable $O(\log(P))$ performance with these sophisticated algorithms. This has only really mattered in very large scale systems where the difference between P and $\log(P)$ is significant. We will see that efficient implementations of these operations exist, for example, in libraries like the popular Message Passing Interface (MPI), to provide high performance algorithms to programmers.

1.3 Why is concurrency considered hard?

The obvious question a programmer should be asking is why is taking advantage of concurrency more difficult than programming sequential programs. Why is concurrency considered to be hard? Concurrency is often treated in computing as a topic that is difficult to express and manage. It is regarded as a constant source of bugs, performance issues, and engineering difficulty. Interestingly, this belief has existed since parallel computers were first proposed. In his 1958 paper, Gill states that:

> Many people with whom the author has spoken have expressed the opinion that programming under such circumstances will be impossibly complicated and will never be worth while. The author feels strongly that this is not so. [40]

Fortunately, Gill was correct in his estimation of the utility of parallel systems and the possibility of programming them. As we will argue in this text, many difficulties programmers face in building programs that use concurrency are a side effect of a legacy of using primitive tools to solve and think about the problem. In reality, we all are quite used to thinking about activities that involve a high degree of concurrency in our everyday, non-computing lives. We simply have not been trained to think about programming in the same way we think about activities as routine as cooking a full meal for dinner. To quote Gill again, he also states this sentiment by pointing out that

> There is therefore nothing new in the basic idea of parallel programming, but only in its application to computers.

1.3.1 Real-world concurrency

Let's consider the activities that we take up in cooking a full meal for ourselves working from a simple recipe. We'll look at how one goes about the process of preparing a simple meal of pasta with home-made marinara sauce.

In the preparation phase, we first sit down and prepare a shopping list of ingredients, such as pasta, herbs, and vegetables. Once the list is prepared, we go to the market and purchase the ingredients. Computationally speaking, this is a sequential process. Enumerating the ingredients is sequential (we write them one by one), as is walking through the market filling our basket for purchase. However, if we happen to be shopping with others, we can split the shopping list up, with each person acquiring a subset of the list which is combined at the checkout for purchase.

The act of splitting the list up is often referred to in computational circles as *forking*. After forking, each person works autonomously to acquire his or her subset of the list without requiring any knowledge of the others during the trip through the store. Once each person has completed gathering his or her portion of the shopping list, they all meet at the checkout to purchase the goods. This is known as *joining*. As each person completes his or her trip through the store, he or she combines his or her subset of the shopping list into the single larger set of items to finally purchase.

After we have purchased the products and returned home to the kitchen, we begin the preparation process of the meal. At this point, concurrency becomes more interesting than the fork-join model that we would utilize during the shopping process. This is due to the existence of *dependencies* in the recipe. Assuming that we are familiar with the recipe, we would likely know that the pasta will take 8 minutes to cook once the water is boiling, while it takes at least 20 minutes to peel the tomatoes for making the sauce. So, we can defer preparing the pasta to prevent overcooking by aiming for it to finish at the same time as the sauce. We achieve this by waiting until the sauce has been cooking for a few minutes before starting the pasta. We also know that the pasta requires boiling water, which takes 5 minutes to prepare from a pot of cold tap water. Deciding the order in which we prepare the components requires us to analyze the dependencies in the recipe process: cooking pasta requires boiled water, pasta sauce requires crushed tomatoes, pasta sauce requires chopped herbs, etc.

This chain of dependencies in this simple cooking problem is illustrated in Figure 1.1. We can observe that the dependencies in the recipe do not form a linear chain of events. The pasta sauce is not dependent on the pasta itself being cooked. The only dependency between the pasta and the sauce is the time of completion — we would like the sauce to be ready precisely when the pasta is, but they have different preparation times. So, we start preparing the ingredients and start subtasks when the time is right. We'll chop the herbs, and start warming the olive oil while we mince the garlic that we'd like to saute before combining with the tomatoes. While the sauce is warming up, we can start the water for the pasta, and start cooking it while the sauce simmers. When the pasta finishes, the sauce will have simmered sufficiently and we'll be ready to eat.

The point of going through this is to show that we are quite used to con-

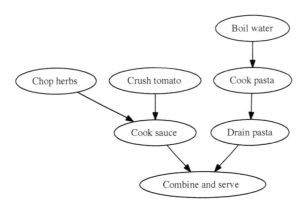

FIGURE 1.1: Dependencies in a simple cooking activity. Arrows point from activities to those that depend on their completion in order to proceed.

currency. Boiling a pot of water at the same time that we prepare part of the meal is just a natural thing we do in the kitchen, and it is an inherently parallel operation. The fact that we can understand, follow, and occasionally write recipes means we understand basic issues that arise in concurrent tasks where dependencies are important and sequential subsets of operations are used. Concurrency really isn't as bad as it is made out to be, *if we have the right methods to express and manage it.*

1.4   Timeliness

As we stated already, the advent of multicore and other novel parallel processors available in consumer hardware has made parallel programming an area demanding significant attention. As of the writing of this text, the state of the art with respect to programming tools for these architectures is primitive at best. Utilizing low core-count processors such as the Intel dual or quad-core processors is best achieved by explicit use of threads in libraries like POSIX threads or compiler extensions like OpenMP. In cases such as the IBM Cell, which represents a potential route that multicore processors will take in the future, programmers are required to work at an even lower level. In the worst case, they must program in assembly or using C library APIs that hide low-level instructions. Furthermore, users must define memory access scheduling explicitly — something that most programmers have assumed to be handled at the hardware level for many years due to the prevalence of cache-based processors.

It is quite likely that as the industry converges on architectures that meet the demands of general purpose computing, tools will arise to assist programmers in using these processors. These tools will provide abstractions above low-level memory access scheduling and the decomposition of instruction streams to various heterogeneous processing elements. We are seeing this today with a rising interest in using graphics co-processors as general purpose processors to achieve higher performance for some algorithms than the traditional main processor alone is capable of.

One area where tools will become available will be in language extensions and the use of novel parallel language constructs. We are seeing a move in this direction with tools like Nvidia's CUDA, the Khronos OpenCL definition, and revisions of existing standards such as OpenMP. The fact that parallel language constructs are designed to allow compilers to easily target parallel systems will mean that we will begin to see technologies from previously obscure languages become integrated with mainstream languages. For this reason, we feel that readers will benefit from learning how these language-level features can be used to construct algorithms in anticipation of tools that will employ them for these new architectures.

1.5 Approach

In this text, we start by laying out for the reader the set of issues that programmers face when building concurrent programs and algorithms. These topics in concurrency are widespread and applicable in many domains, ranging from large scale scientific programming to distributed internet-based systems and even into the design of operating systems. Familiarity with these topics and the consequences they have in design decisions and methods is vital for any programmer about to embark on any task where concurrency is a component. In order to ground this topic in areas with which programmers are already familiar, we will discuss the relationship that concurrency has to traditional sequential models used frequently in everyday programming tasks. Most importantly, the reader should understand precisely why traditional programming techniques and models are limited in the face of concurrency.

Before delving into the description of the high-level language constructs that programmers should begin to think about algorithms in terms of, we will discuss the current state of the art in concurrent programming. This discussion will not only serve as a summary of the present, but will expose the flaws and limitations in the current common practices that are addressed by higher-level language constructs.

In introducing high-level parallel programming techniques, we will also discuss notable milestones in the hardware evolution that occurred over the past forty years that drove the developments in linguistic techniques. It has repeatedly been the case that new hardware has been preceded by similar hardware introduced decades earlier. Multimedia extensions in consumer level hardware derive their lineage from vector processing architectures from the late 1970s and 1980s. The first dual and quad core multicore processors that arrived a few years ago are essentially miniaturized versions of shared memory parallel systems of the 1980s and 1990s. More recently, these multicore processors have taken on what can best be described as a "hybrid" appearance, borrowing concepts tested in many different types of architectures. We will discuss this historical evolution of hardware, the corresponding high-level techniques that were invented to deal with it, and the connection that exists to modern systems. Much of our discussion will focus on shared memory multiprocessors with uniform memory access properties. Many of the concepts discussed here are applicable to nonuniform memory access shared memory machines and distributed memory systems, but we will not address the nuances and complications that these other architectures bring to the topic.

Finally, we will spend the remainder of the text on a discussion of the various high-level techniques that have been invented and their application to a variety of algorithms. In recent years, a very useful text by Mattson et al. emerged to educate programmers on patterns in parallel programming [69]. This text is similar to the influential text on design patterns by Gamma et al., in that it provides a conceptual framework in which to think about the different aspects of parallel algorithm design and implementation [37]. Our examples developed here to demonstrate and discuss high-level language techniques will be cast in the context of patterns used in Mattson's text.

1.5.1 Intended audience

This book is intended to be used by undergraduate students who are familiar with algorithms and programming but unfamiliar with concurrency. It is intended to lay out the basic concepts and terminology that underly concurrent programming related to correctness and performance. Instead of explaining the connection of these concepts to the implementation of algorithms in the context of library-based implementations of concurrency, we instead explain the topics in terms of the syntax and semantics of programming languages that support concurrency. The reason for this is two-fold:

1. Many existing texts do a fine job explaining libraries such as POSIX threads, Windows threads, Intel's Threading Building Blocks, and compiler directives and runtimes such as OpenMP.

2. There is a growing sense in the parallel programming community that with the advent of multicore and the now ubiquitous deployment of

small-scale parallelism everywhere there will be a trend towards concurrency as a language design feature in languages of the future.

As such, it is useful for students to understand concurrency not as a library-like add on to programs, but as a semantic and syntactic topic of equal importance and treatment as traditional abstractions that are present in the programming languages of today.

1.5.2 Acknowledgments

This book is the result of many years of work on the part of the authors in the parallel computing community, and is heavily influenced by our experiences working with countless colleagues and collaborators. We would like to specifically thank David Bader, Aran Clauson, Geoffrey Hulette, Karen Sottile, and our anonymous reviewers for their invaluable feedback while writing this book. We also are very grateful for the support we have received from our families who graciously allowed many evenings and weekends to be consumed while working on this project. Your support was invaluable.

Web site

Source code examples, lecture notes, and errata can be found at the Web site for this book. We will also keep a relatively up-to-date list of links to compilers for languages discussed in this book. The URL for the Web page is:

```
http://www.parlang.com/
```

1.6   Exercises

1. Describe an activity in real life (other than the cooking example in the text) that you routinely perform that is inherently concurrent. How do you ensure that the concurrent parts of the activity are coordinated correctly? How do you handle contention for finite resources?

2. List a set of hardware components that are a part of an average computer that require software coordination and management, such as by an operating system, due to their parallel operation.

3. Using online resources such as manufacturer documentation or other information sources, write a short description of a current hardware platform that supports parallelism. This can include multicore processors, graphics processing units, and specialized processors such as the Cell.

4. Where in the design and implementation of an operating system do issues of concurrency appear?

5. Why is concurrency important to control in a database? Can you think of an instance where improper management of clients can result in problems for a database beyond performance degradation, such as data corruption?

6. Where does concurrency appear in the typical activities that a Web server performs when interacting with clients?

7. Beginning with all of the parts of a bicycle laid out in an unassembled state, draw a chart (similar to Figure 1.1) that shows the dependencies between the components, defining which must be assembled before others. Assume a simple bike with the following parts: wheels, tires, chain, pedals, seat, frame, handlebars.

Chapter 2

Concepts in Concurrency

Objectives:

- Establish terminology for discussing concurrent programs and systems.
- Discuss dependencies and their role in concurrent programs.
- Define important fundamental concepts, such as atomicity, mutual exclusion and consistency.

When learning programming for sequential machines, we all went through the process of understanding what it means to construct a program. The basic concept of building solutions to problems within a logical framework formed the basis for programming. Programming is the act of codifying these logic-based solutions in a form that can be automatically translated into the basic instructions that the computer executes. To do so, we learned one of a variety of programming languages.

In doing so, we learned about primitive building blocks of programs, such as data structures, statements and control flow primitives. We also typically encountered a set of fundamental algorithms, such as those for searching and sorting. This entailed learning the structure and design of these algorithms, syntax for expressing them in a programming language, and the issues that arise in designing (and more often, debugging) programs. Along the way, convenient abstractions such as loops and arrays were encountered, and the corresponding problems that can arise if one is not careful in their application (such as errors in indexing).

As programs got more complex, we began to employ pointers and indirection to build powerful abstract data structures, and were exposed to what can go wrong when these pointers aren't maintained correctly. Most programmers have at some point in their careers spent a nontrivial amount of time tracking down the source of a segmentation fault or null pointer exception. One of the driving factors behind high-level languages as an alternative to working with lower-level languages like C is that programmers are insulated from these issues. If you can't directly manipulate pointers, the likelihood of using them incorrectly is drastically reduced. Convenient higher-level language constructs abstracted above this lower-level manipulation of raw memory layouts and contents, and removed many of the tedious details required when building correct and efficient programs by hand.

Practitioners in programming rapidly learn that it is a multifaceted activity, involving gaining and mastering an understanding of:

- How to codify abstract algorithms in a human readable form within a logical framework.

- How to express the algorithm in the syntax of a given programming language.

- How to identify and remedy correctness and performance issues.

Much of the latter part of this text will focus on the first two points in the context of parallel programming: the translation of abstract algorithms into the logical framework provided by a programming language, and the syntactic tools and semantics provided by the language for expressing them. This chapter and the next focus on the third point, which as many working programmers know from experience is a crucial area that requires significant thought and practice even when the act of writing code becomes second nature. Furthermore, the design aspect of programming (an activity under the first point) is intimately related to the considerations discussed in this chapter.

The operations from which sequential programs are constructed have some assumed behavior associated with them. For example, programmers can assume that from their perspective one and only one operation will occur on their data at any given time, and the order of the sequence of operations that does occur can be inferred directly from the original source code. Even in the presence of compiler optimizations or out-of-order execution in the processor itself, the execution can be expected to conform to the specification laid out in the original high-level source code. A contract in the form of a *language definition*, or *standard*, exists between the programmer and language designer that defines a set of assumptions a programmer can make about how the computer will behave in response to a specific set of instructions.

Unfortunately, most languages do not take concurrency into account. Concurrent programming complicates matters by invalidating this familiar assumption — one cannot infer relative ordering of operations across a set of concurrent streams of instructions in all but the most trivial or contrived (and thus, unrealistic) circumstances. Similarly, one cannot assume that data visible to multiple threads will not change from the perspective of an individual thread of execution during complex operation sequences. To write correct concurrent programs, programmers must learn the basic concepts of concurrency just like those learned for sequential programming. In sequential programming, the core concept underlying all constructs was logical correctness and operation sequencing. In the concurrent world, the core concepts are not focused solely on logic, but instead on an understanding of the effect of different orderings and interleavings of operations on the logical basis of a program.

These initial chapters introduce the reader to the problems that can arise in this area in the form of non-determinism, and the conceptual programming constructs that exist to manage and control non-determinism so that program correctness can be inferred while benefiting from the performance increases provided by supporting the execution of concurrent instruction streams. We will also see that a whole new class of correctness and performance problems arise if the mechanisms to control the effect of non-determinism are not used properly. Many of these topics are relevant in the design of network servers, operating systems, distributed systems and parallel programs and may be familiar to readers who have studied one or more of these areas. Understanding these concepts is important as a basis to understanding the language constructs introduced later when we discuss languages with intrinsically parallel features.

2.1 Terminology

Our first task will be to establish a terminology for discussing parallel and concurrent programming concepts. The terminology will be necessary from this point forward to form a common language of discussion for the topics of this text.

2.1.1 Units of execution

In discussing concurrency and parallelism, one must have terminology to refer to the distinct execution units of the program that actually execute concurrently with each other. The most common terms for execution units are *thread* and *process*, and they are usually introduced in the context of the design and construction of operating systems (for example, see Silberschatz [87]).

As we know, a sequential program executes by changing the *program counter* (PC) that refers to the current position in the program (at the assembly level) that is executing. Branches and looping constructs allow the PC to change in more interesting ways than simply regular increasing increments. Similarly, the program is assumed to have a set of values stored in CPU registers, and some amount of memory in the main store available for it to access. The remainder of this context is the operating system state associated with the program, such as I/O handles to files or sockets. If one executes the same program at the same time on the machine, each copy of the program will have a distinct PC, set of registers, allocated regions of memory and operating sys-

Components of a single threaded
process.

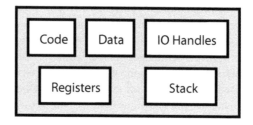

FIGURE 2.1:   An operating system process and its components.

tem state. Programs, unless explicitly coded to do so, cannot see the state
of other programs and can safely treat of the machine as theirs exclusively.
The only artifact of sharing it with other programs would most likely be observed as performance degradation due to contention for resources. This self
contained encapsulation of the environment and control state of the program
is what we refer to as the execution unit or execution context.

In modern operating systems, these encapsulations of data and control state
are referred to as *processes*. For the most part, they exist to enable multitasking operating systems by providing the necessary abstraction of the machine
from the program, and also some measure of protection such that poorly be-
having programs cannot easily disrupt or corrupt others.

One of the simplest methods to implement concurrent programs is by creating a set of processes that have a means to communicate with each other
by exchanging data. This is precisely how many concurrent systems work. It
is safe to assume many readers familiar with C will have seen the `fork()` call
at one point, which allows a single program to create (or "spawn") multiple
instances of itself as children that can execute concurrently. Simple server
applications are built this way, where each forked child process interacts with
one client. A parent process will be run, and as jobs arrive (such as requests
to an HTTP server), child processes are created by forking off instances of
the parent that will service the operations associated with each incoming task
that arrives, leaving the parent to wait to service new incoming tasks.

The term *fork* typically has the same high-level meaning in all cases, but can
be interpreted slightly differently depending on the context. When referring
to the flow of control within a program, if this flow of control splits into
two concurrently executing streams of execution, we would say that the flow
of control had forked. In this usage, the implementation of the concurrent
control flow isn't specified — it could be via threads or processes. On the
other hand, the `fork()` system call typically has a specific meaning that is

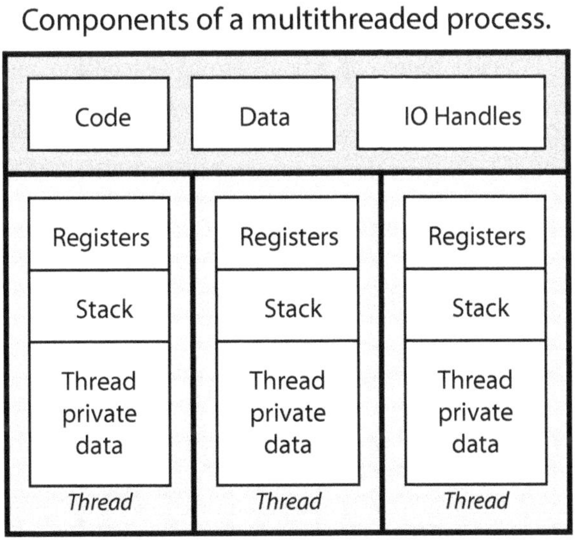

Components of a multithreaded process.

FIGURE 2.2: A multithreaded process, showing the separation of process-level from thread-level information.

process oriented (even though the actual implementation may in fact be based on kernel threads for efficiency purposes).

This model of creating multiple processes relieves the programmer of explicitly managing and servicing multiple connections within a single program, deferring this to the operating system multitasking layer. Programs that do not do this via multiple processes are possible to implement (such as through the use of the select() call), but tend to be more tedious to construct as the programmer must manually manage the state of the set of concurrent clients accessing the program. In a manner similar to using the fork() operation, remote procedure call (RPC) and interprocess communication (IPC) libraries exist to facilitate interactions between distinct processes necessary to form concurrent programs out of them.

The term "process" unfortunately carries with it too much operating system oriented baggage, and in fact, processes themselves tend to have a great deal of baggage in the form of performance overhead in their practical use for building concurrent programs. This is where threads come into play.

Processes require allocation of resources both to support the process itself, and within the operating system for tracking resources and making scheduling decisions that support multiprocessing. As such, creation of a process is a relatively heavy-weight operation — for sufficiently simple operations, the cost of creating and destroying a process may be far greater than the actual work that it was to perform. For this reason, lighter-weight entities known as *threads* were invented. As illustrated in Figure 2.2, a thread is a component of a process. Threads do not require the more intrusive setup and tear-down overhead of a process, and instead exist within a process that already has been created. Threads represent a very simple execution unit that differs from processes in that:

- Threads share a single memory space with peers within the same process.

- Threads share I/O resources within a process.

- Threads have independent register state and stacks.

- Threads may have private, thread-local memory.

Threads are important for concurrent programming. They are simpler than processes and can require significantly less overhead to create or destroy. The amount of overhead depends on the method of thread implementation by the underlying system. Furthermore, they facilitate easier concurrent programming for programs that require interactions between the concurrently executing units because threads share the common process context from which they were created. The fact that all threads created by a single process share some common, process-level state with each other makes it easier to build programs that require multiple execution units to take advantage of each other's state.

Instead of tedious functions for sending data back and forth between separate threads, they can simply refer to the state of others through standard variables. A set of threads can interact with each other by manipulating this shared process level state that the operating system provides. A shared state (often referred to as a *shared memory*) allows code to directly address data, as opposed to the methods provided by RPC and IPC systems in which sharing requires explicit copying and memory management between disjoint memory spaces. Shared memory programming in threaded code is often supported by something as simple as managing the scope of shared variables. A shared variable must be defined outside of the thread body but still lexically in its scope. Note that when we refer to shared memory in the context of this text, we are *not* referring to older shared memory systems for interprocess communication (such as System V IPC) in which processes can tag process-local data to be visible by others. While some of the same correctness issues can arise in that case, interprocess shared memory is often a library level technique that is not introduced via programming language constructs, and thus is outside the scope of this text.

Shared memory programs are arguably easier to write that those utilizing separate processes, as they require no additional API for building programs out of concurrently executing, interacting parts. Unfortunately, shared memory, multithreaded programming opens the door to many new types of correctness issues that are otherwise not present in a sequential program.

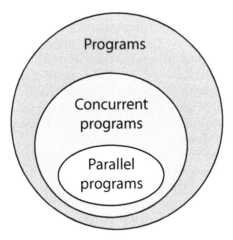

FIGURE 2.3: Parallel execution is a special case of concurrent execution, which itself is a specific case of general program execution.

2.1.2 Parallelism versus concurrency

We briefly introduced the difference between parallelism and concurrency in the introduction, and will now study the difference in more detail. The terms parallel and concurrent are often used in what appears to be an interchangeable way, and they are used in this text often. Why do we have two terms for what appears to be the same concept?

One can mistake them for equivalent terms, as often "parallel computing" is used to mean the same thing as "concurrent computing," and vice versa. This is incorrect, although for subtle reasons. The simplest way to distinguish them is that one focuses on an abstract view of how multiple streams of instructions execute over time ("concurrency"), while the other focuses on how they actually execute relative to each other in time ("parallelism"). We will show some simple examples to illustrate this distinction. In fact, a parallel program is simply a specific execution instance of a concurrent program as shown in Figure 2.3.

Consider two streams of operations that are, for the intents of this discussion, independent and unrelated. For example, a user application and an operating system daemon. The beauty of modern multitasking operating systems is that an abstraction is presented to the user that gives the appearance of these two tasks executing at the same time — they are *concurrent*. On the other hand, on most single processor systems, they are actually executing one at a time by interleaving instructions from each stream so that each is allowed to progress a small amount in a relatively short period of time. The speed of processors makes this interleaving give the appearance of the processes running at the same time, when in fact they are not. Of course, this simplified view of the computer ignores the fact that operations such as I/O can occur for one stream in hardware outside of the CPU while the other stream executes on the CPU. This is, in fact, a form of parallelism.

We define a **concurrent** program as one in which multiple streams of instructions are active at the same time. One or more of the streams is available to make progress in a single unit of time. The key to differentiating parallelism from concurrency is the fact that through time-slicing or multitasking one can give the illusion of simultaneous execution when in fact only one stream makes progress at any given time.

In systems where we have multiple processing units that can perform operations at the exact same time, we are able to have instruction streams that execute in *parallel*. The term parallel refers to the fact that each stream not only has an abstract timeline that executes concurrently with others, but these timelines are in reality occurring simultaneously instead of as an illusion of simultaneous execution based on interleaving within a single timeline. A concurrent system only ensures that two sequences of operations may appear to happen at the same time when examined in a coarse time scale (such as a user watching the execution) even though they may do so only as an illusion through rapid interleaving at the hardware level, while a parallel system may actually execute them at the same time in reality.

Even single CPU systems have some form of parallelism present, as referred to above in the context of I/O. When a program requires I/O of some form, the interface to the storage or communication peripheral is used to perform the often high latency operation. During this time, while the I/O operation is executing in the hard drive or network hardware, other processes may execute on the main processor to get useful work done. This is a crude form of parallelism, and although very useful for multitasking workloads, is of little value within single programs that are more computationally focused. Interestingly, the parallel execution of I/O with computation was one of the first accepted uses of parallelism in early computing systems.

Our definition of a **parallel** program is an instance of a concurrent program that executes in the presence of multiple hardware units that will guarantee that two or more instruction streams will make progress in a single unit of time. The differentiating factor from a concurrent program executing via time-slicing is that in a parallel program, at least two streams make progress at a given time. This is most often a direct consequence of the presence of multiple hardware units that can support simultaneous execution of distinct streams.

In both cases, the difference between concurrency and parallelism is a consequence of the context in which a program is executed. A single program that uses multiple threads of execution may be run in both a sequential environment that employs time-slicing and a parallel environment with multiple processors available. The notion of concurrency captures both potential execution environments, while parallelism is more specific. The issues that we address in this text can arise in both cases though.

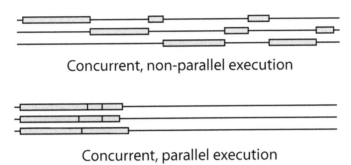

Concurrent, non-parallel execution

Concurrent, parallel execution

FIGURE 2.4: An illustration of progress made by three threads of execution executing concurrently and sharing a single resource, and executing in parallel with their own exclusive resources.

General purpose parallel capabilities are often realized by replicating the computational elements of the system so that each stream of instructions could proceed without interfering with the others. This has been realized in various forms, either vector processing units, multicore CPUs, multi-processor systems, and large scale supercomputers composed of hundreds or thousands of processors. We will discuss later how to differentiate these different methods for implementing parallelism, which impacts both hardware and software structure.

Now that we've distinguished what concurrency is relative to parallelism, we will proceed in our discussion of concurrent programming. Due to the fact that the user often has little to no control over how concurrent instruction streams are interleaved on shared hardware resources, the correctness issues and corresponding solutions that arise in parallel computing are identical to those in concurrent programming. Thus, for general applicability, we will concern ourselves with creating programs that are composed of concurrently executing parts, regardless of whether or not they execute by interleaving or temporal parallelism. The core of this topic relates to the issue of interference alluded to earlier. How do concurrent or parallel instruction streams interfere with each other, and what are the ramifications to programmers who wish to create reliable, correct programs?

2.1.3 Dependencies and parallelism

A fundamental part of thinking about parallel or concurrent programs is identifying parallelism that can be exploited in a problem or program. This task is independent of how one actually implements the program and whether or not it executes in parallel or as a set of time-sliced concurrent execution units. The key to identifying potential parallelism in a problem is identifying *dependencies* within it.

A dependency is something, either a state of data or a control state, which must be reached before a part of a program can execute. For example, consider the following sequential code.

```
x = 6;
y = 7;
z = x+y;
```

The first two statements depend on no others, as they correspond to assigning constant values to two variables. The third statement does have a dependency. The value assigned to z cannot be computed and stored until both of the statements before it are complete. We would say that this third statement *depends on* the first and second statements.

Dependencies are important because they impose an ordering on when statements can execute relative to each other. If a statement depends on another, then it cannot execute until those it depends on have completed. Chains of dependencies impose a sequential structure on parts of the program — there is no way to execute two statements in parallel and obtain meaningful results if one depends on the value of the other. If this is attempted, we would either see the dependent statement wait until the first completes (serializing the code, which defeats the point of parallelism), or the dependent statement would execute based on incorrect, incomplete, or outdated inputs if the first statement hadn't completed yet. Later, we will see that the dependency relationships within a program have important impacts on performance due to Amdahl's Law.

Fortunately, very few programs exist in which the chain of dependencies forms a single linear sequence. Even in the basic example above, we have two statements that do not depend on each other (x=6; and y=7;), so the order in which they execute doesn't matter as long as the third statement waits until both have completed. Dependencies are not strictly limited to statements within a program. We can consider coarser grained parts of a program, such as subroutines, object methods, blocks of code, or even complex subprograms involving many routines and objects. Any meaningful decomposition of a program into a set of tasks can be analyzed to understand the dependency relationships between them.

Bernstein introduced and formalized this notion of dependencies in the context of parallel programs in 1966 by laying out what have come to be known as *Bernstein's conditions* [11]. Bernstein's conditions are concerned with dependencies based on memory locations that a statement or subprogram uses for input or for output. In the original paper introducing the conditions, the set of locations was broken into four classes based on whether or not a subprogram read from them or wrote to them, and whether or not a read was followed by a write, or vice versa. The conditions are often generalized in a slightly simpler presentation than the original paper provided.

TABLE 2.1: Dependencies between three simple statements.

Statement (S)	$IN(S)$	$OUT(S)$
x=6	$\emptyset$	$\{x\}$
y=7	$\emptyset$	$\{y\}$
z=x+y	$\{x, y\}$	$\{z\}$

Consider a statement or subprogram P. Let $IN(P)$ be the set of memory locations (including registers) or variables that P uses as input by reading from them. Let $OUT(P)$ be the set of memory locations or variables that P uses as output by writing to them. In our simple example above, we can define these sets for each line as shown in Table 2.1 (the symbol $\emptyset$ corresponds to the empty set).

Bernstein's conditions state that given two subprograms P_1 and P_2, the sequential execution of P_1 and P_2 is equivalent to their execution in parallel if the following conditions hold.

- **BC1:** $OUT(P_1) \cap OUT(P_2) = \emptyset$

- **BC2:** $IN(P_1) \cap OUT(P_2) = \emptyset$

- **BC3:** $IN(P_2) \cap OUT(P_1) = \emptyset$

The first condition (BC1) states that the output sets of each subprogram are independent. This is important because if they are not independent, then there is some location into which both subprograms will be writing. The result of this would be that the final value that the location would contain would be dependent on the order in which the two subprograms P_1 and P_2 complete. Parallel execution does not impose any ordering on which completes first, so any nonempty intersection in this output set would lead to unpredictable results if the two subprograms were executed in parallel. The second and third conditions (BC2 and BC3) are related to input/output dependencies between the subprograms. If there is any location that one subprogram reads from that the other writes to, then a dependency exists that dictates which must execute first in order to make input data available to the other.

As we will see later, concurrency control primitives exist to coordinate the execution of concurrently executing tasks such that these conditions hold. We will also discuss correctness problems that may result when they are violated. Our discussion of languages and computational models for parallelism will also show how the use of dependency properties of a program can be used to motivate a family of languages and hardware platforms based on the dataflow model.

2.1.4   Shared versus distributed memory

Parallel computers are most frequently split into two coarse groupings — those based on shared and distributed memory. In this text, many of the language techniques are based on threads which share memory, so they naturally fit into machines that adopt the shared memory model. Those that employ message passing can be adapted to work on both shared or distributed memory. What then distinguishes parallel machines with shared versus distributed memories?

In a shared memory system, multiple processing units can directly address locations in a single shared address space. If two processors that share memory reference an address, they are referring to the same physical location in memory. Shared memory multiprocessors have existed for a long time in the server market and high-end workstations. More recently, multicore processors have brought this architectural design to the desktop and laptop. Within a multicore computer, each processor core shares the main memory with the others. Therefore multicore computers at the current time are instances of shared memory parallel computers.

The distributed memory model is also a popular architectural model that is most frequently encountered in networked or distributed environments, although it can occur in more tightly coupled designs in specialized situations. In a distributed memory system, there exist at least two processing elements that cannot directly address memory that is directly addressable by the other.

For example, if we take two computers and connect them by a network, programs on each individual computer can directly address memory physically attached to the processor that they execute on. They cannot directly address the memory that is physically connected to the computer on the other end of the network connection. If they wish to read from this memory, they do so by sending a message over the network so that the computer that "owns" the memory can access it on their behalf and send the data contained in the memory over the network to the requester. Similarly, writes must be performed by sending the data to be written to the owner of the memory so that they can perform the actual update to physical memory.

When we consider the whole set of processors in a parallel machine, the terminology makes more sense. If all processors in the machine can address the entire memory space directly, we would say that they share it. On the other hand, if each processor can access only a subset of the physical memory, then we would say that the overall memory space is distributed amongst the processors.

2.2   Concepts

There are a small set of abstract concepts that underly all concurrent programming constructs. These concepts are used to ensure program correctness by providing programmers with behavioral guarantees and are not difficult to understand, but are often overlooked by programmers who spend most of their time writing sequential programs. When writing programs that execute as a set of concurrent interacting processes, programmers must not think about code purely as a set of operations that operate in sequence. Concurrent programming requires one to consider not only the concurrent sequences of operations in isolation from each other, but how their individual execution is impacted by, or has an impact upon, the execution of other sequences of operations. Interesting, and at times problematic things occur when processes executing in parallel interact with one another. The concepts discussed here cover how one must think about such interacting processes and the issues that arise as they interfere with each other.

Historically speaking, many of these concepts were defined (and the constructs invented to deal with them from a correctness point of view) in the late 1960s through the 1970s. Four computer scientists can be credited with laying much of the foundation of parallel and concurrent programming: Edsger Dijkstra, C.A.R. (Tony) Hoare, Per Brinch Hansen, and Leslie Lamport. These four were present at the dawn of concurrent computing, when machines reached a level of sophistication in the 1960s where new ideas and points of view became feasible. The primary area where concurrency was of interest was in operating system design. As processing speeds increased, and the gap between this speed and I/O devices grew, it was increasingly difficult to keep machines fully utilized. Utilization refers to the amount of time a machine spends doing useful work over a fixed time interval. For example, if a computer spends 5 seconds doing I/O (with the processor waiting), and then 5 seconds doing computation in a 10 second interval, we would say that the machine has a utilization of 50%.

From an economic point of view, this is wasteful — why not use the time spent idle achieving useful work on another task instead of letting valuable computer resources go to waste? Don't forget that early on, a single computer was a very large expense to any organization, so every compute cycle was precious. This is precisely the problem that motivated operating system designers: how can machines be used by multiple tasks such that they can be interleaved with each other to fill in the idle time that arises while they execute?

In addition to the gap in I/O versus processor performance, machine hardware reached a point, from an engineering and economic perspective, where it was feasible to build computers that contained multiple, independent processing units. Early machines included dedicated co-processors for assisting with I/O, and in some cases computation. How to program these early instances of parallel systems required investigation. Later, another hardware development revealed more instances of concurrency with the advent of networking. This made it possible to couple multiple computers together to act in concert to perform operations more complex than any individual system was capable of alone.

Little has changed since then in the fundamental nature of parallel and concurrent systems — the hardware has simply become orders of magnitude more capable in terms of speed and capacity. Fundamentally, what makes a parallel or concurrent system, and the resulting properties of that system, has not changed. The paradigm shifted in the 1960s, and the field of computing has simply refined it as technology changed and people found new, creative ways to utilize it. Even in the modern world of multicore computing, the fundamental nature of the systems hasn't changed — it has simply reached a point where an entire machine room from earlier decades now fits on a single silicon die.

2.2.1 Atomicity

Nearly all programs that are not side effect free (such as non-purely functional programs) operate by reading, operating on, and modifying data stored in some form of memory. Even in a purely functional programming model, the underlying implementation of the programming abstraction that is emitted by the language compiler is based on this memory modification (or "side effect") based model. This model is often referred to as the von Neumann model of computation. Programs are constructed as a sequence of operations with the primary purpose of modifying the state of values stored in memory. One of the fundamental properties of programs in this model is that they induce effects — by executing them, something about the underlying system changes. Most often, effects have the form of changes to the state of memory. This model is based on executing programs by the following sequence of fundamental operations:

1. Fetch value from main memory into a processor register.

2. Operate on the register, store the result in a register.

3. Store the value from the register back into main memory.

Furthermore, the programs themselves reside in memory, and execute by incrementing a program counter representing the current position within the program that is executing. This model has proven to be exceptionally flexible, as any owner of a modern computing device can attest to. As we will see though, the effect of concurrency on this model of computing leads to a number of interesting issues.

Listing 2.1: A simple C function incrementing an element of an array.

```
void fun() {
  int x[100];
  x[42]++;
}
```

To explore these issues, we should first examine how programs in the von Neumann model are constructed. Programs are simply sequences of operations that either manipulate data or modify the program counter in order to represent sequences of execution beyond simple straight line code. Each operation does something simple and the aggregate effect of well coordinated sequences of these operations implements the desired program. These operations range from performing basic arithmetic movement of data within the computer, manipulating the value of the program counter based on the conditional interpretation of a value stored in the program state, to invoking self-contained and parameterized sequences of operations in the form of subroutines. The key observation is that programmers build programs out of operations that are, from their point of view, *atomic*. An **atomic operation** is an operation that cannot be split into multiple smaller operations, and will execute as if it was a single operation with exclusive control over the portions of the system that it requires during its execution.

The choice of the term *atomic* is quite intentional. The word *atom* derives from the Greek word *atomos* (ατομος), meaning indivisible. The interpretation of this concept is subtle. An atomic operation is intended to be treated as indivisible from the perspective of the programmer, even though it may be composed of a sequence of finer grained operations. The point of treating something as atomic is not that is in fact indivisible, but that if it is ever divided into its constituent parts, any manipulation of this division could result in an operation that no longer corresponds with the original aggregate atomic operation.

To illustrate the cause and potential effects due to atomic versus non-atomic operations, consider the simple C function in Listing 2.1 that increments a single element from a locally stack allocated array.

Ignoring the fact that the array is not initialized with any specific values, we can look at how the increment operation actually is executed at the assembly level as shown in Listing 2.2. Examining the x86 assembly language emitted by the C compiler that corresponds to this simple function, we see that the increment operation "++" is not guaranteed to be a single operation in its implementation on the machine.[1] It is in fact a sequence of assembly

---

[1] We demonstrate the non-atomic implementation of ++ for illustrative purposes here. Often a compiler implements this operation in an atomic fashion.

Listing 2.2: The x86 assembly that corresponds to Listing 2.1.

```
1                  .text
2       .globl  _fun
3       _fun:
4                  pushl    %ebp
5                  movl     %esp, %ebp
6                  subl     $408, %esp
7                  movl     -240(%ebp), %eax
8                  addl     $1, %eax
9                  movl     %eax, -240(%ebp)
10                 leave
11                 ret
12                 .subsections_via_symbols
```

operations.

The key lines in the assembly are lines 7 through 9. On line 7, the specific entry in the array is loaded into a register (the value -240 is the offset into the stack frame for the stack allocated array — this is unimportant.) On line 8, the value is incremented by one, and line 9 commits the value from the register back to main memory. Even though line 3 from the original C code is atomic from the programmer's perspective, it is not actually atomic in its implementation. The programmer's assumption of atomicity means that this sequence of operations corresponding to the ++ operator must not be divided in a way that will deviate from the result achieved by their contiguous execution.

In sequential programs, atomicity is often ignored and taken for granted, as it often doesn't matter how many distinct operations occur in sequence to execute the algorithm correctly on a sequential computer. The only time when the specific sequence of operations is of interest to the programmer is in performance tuning where one wishes to minimize the number of wasteful operations executed. More often than not, the programmer has simply been concerned with ensuring that the ordering is correct and that higher level invariants hold during the execution related to algorithmic correctness.

In concurrent programs, atomicity becomes a significant issue when multiple streams of control have access to common data. This is because a sequence of operations itself is not necessarily atomic. The sequence is composed of a set of operations that occur in some assumed order, but the sequence does not prevent the underlying hardware and operating system from performing other operations during their execution. The only guarantee the system provides is that this sequence executes in the correct order. Other operations may occur during the execution of this sequence that are related to some other task.

When two concurrent instruction sequences execute, it is possible that an operation from the second sequence will occur at some point during the exe-

Listing 2.3: Code snippet to run in parallel.

```
shared int counter;
private int x;
x = counter;
x = x+1;
counter = x;
```

cution of the first due to interleaving of the operation sequences in time. The order of each will be preserved relative to themselves, but their order relative to each other is undefined. When these two processes share data, these interleaved operations can result in one process reading data that is partially updated by the other, or modifying data that the other has assumed to be in some known state. Most importantly, no guarantees are made with respect to how they manipulate the memory state accessible by both. This can lead to significant problems.

Example: Incrementing a shared counter

To illustrate sequences of operations where an atomicity assumption may hold, consider two processes that have access to a shared counter variable. The programmer assumes that the operation of incrementing the counter itself is semantically atomic, even though the operation itself consists of reading the shared counter into a private local variable (possibly a register), incrementing it, and storing the new value back into the shared location. This is illustrated earlier in Listing 2.1 for the "++" operator with the resulting assembly in Listing 2.2. We write the sequence of operations that actually implement the increment operator in pseudo-code in Listing 2.3. The increment operator is implemented as a sequence of three operations. Atomicity is important for this operator, as we must guarantee that each increment occurs without being lost due to interactions with other processes. Unexpected execution behavior that results in incorrect results in the counter can cause problems for the program relying on the value of the counter.

What can go wrong with this example? Notice that nothing about the instruction streams makes them aware of each other and the location in the code where each lies. Thus, the programs may execute in many different interleaved orderings. If we are lucky, the interleaving shown in Figure 2.5 will occur and the counter will have the expected value. An ordering that can lead to incorrect results is shown in Figure 2.6. Assume that the first process executes statement 3, and before proceeding further, the second process executes statement 3. Both processes then proceed with the original counter value, incrementing it by one, and storing the result back. The second process is unaware that the first had already entered the counter increment section, and it increments a value that is outdated from the point of view of the programmer.

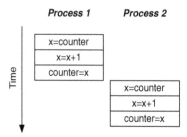

FIGURE 2.5: Two processes executing a non-atomic counter increment without interleaving.

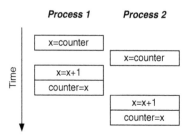

FIGURE 2.6: Two processes executing a non-atomic counter increment with erroneous interleaving.

If the counter starts with a value of n, after the case with no interleaving, the result after both increments have occurred will be the expected value of $n+2$. If interleaving occurs as shown, the result will be $n+1$, which is obviously incorrect given that two counter increments were expected to have occurred!

This sort of result due to uncontrolled interleaving can be devastating for systems where complex sequences of operations must be assumed to be atomic in order to guarantee consistency of the shared memory state. Process ID allocation in operating systems, database key allocation, or simply manipulation of shared counters are real world examples where this must be taken into consideration. How do we ensure atomicity when it is necessary? Doing so requires the use of synchronization between concurrently executing threads. One mechanism by which synchronization can be used to achieve atomicity is by *mutual exclusion* and the identification of *critical sections* of code.

2.2.2   Mutual exclusion and critical sections

As we saw in discussing atomicity, often algorithms require sequences of operations to all execute *as if* they were a single operation in which the memory state they access is not modified by any external source while the sequence executes. These complex, semantically atomic operations are in fact non-atomic sequences of primitive operations. Atomicity is guaranteed only within each primitive operation and not between them. This means that additional steps must be taken to guarantee atomicity of the sequence in the presence of other processes with access to shared data. These sequences of code, where atomicity is assumed at a higher level of abstraction, are often referred to as *critical sections* or *critical regions*.

A critical section is a region of code where the programmer must assume that one and only one sequence of operations has access to read or modify shared data accessed by that region. It appears in many areas of programming, far beyond that of building parallel programs. Two familiar and important examples that you have likely encountered previously include:

- **Databases:** Say a database update will modify the address of a person in a table. This update includes modifying the street address, city, state, and other information. If the full address update is not considered atomic, it is possible that another process will query the database during the update and end up with a record that is inconsistent. The street and city may correspond to the new address, while the state may correspond to the old.

- **Operating systems:** Nearly all modern operating systems associate a unique identifier (often an integer) to each process that executes on the computer. It is extremely important that these identifiers be unique, as dreadful consequences can arise if they are not. Imagine using a system where a simple "kill" command could impact other processes than the one the operation was intended for. The results would be disastrous!

Readers with experience working with databases will be very familiar with the issues that arise due to the need for atomic operations. Database transactions are a means for implementing atomicity for sequences of database operations. We will discuss this in greater detail later when looking at software transactional memory, a database-inspired technique of some interest to the parallel programming community.

Mutual exclusion is the concept of ensuring that one and only one process can execute operations that must be considered atomic for reasons of correctness, and the corresponding protection of data that must be kept consistent during the critical region. Often this is realized in programs by the defining the beginning and end of such regions of code, with possible annotations indicating which data elements within the region must be protected. The reason why only specific data elements may be called out as in need of protection is to prevent unnecessary restrictions on data that is in scope during the critical region, but whose protection is irrelevant to the correctness of the critical region itself.

Implementing mutual exclusion typically involves the acquisition of permission by a process to enter the critical region, and notification that the region is no longer occupied when exiting the critical region. How one implements this notion of permission to enter and notification of completion is very dependent on the underlying concurrency support. Most often this is implemented with some synchronization mechanism, such as a semaphore, lock, or monitor. These are all discussed in detail in the next chapter.

If the concept of mutual exclusion is unclear, consider the non-computational example of a small public restroom. Obviously one wishes to avoid multiple occupants, so there must be a method in place to ensure that one and only one person is occupying the "resource" at any given time. A simple lock (with the occasional corresponding external sign that switches between "occupied" and "available") maintains the mutually exclusive access to the restroom. In concurrent programming, we often wish to exercise the same protocol for controlling access to a sensitive or critical resource.

2.2.3   Coherence and consistency

It is often useful from the perspective of the programmer to reason about how concurrent processes interact relative to how they would behave if they were reduced to an interleaved sequential stream of operations, as we saw earlier in Figure 2.6. Lamport introduced this way of thinking about concurrent versus sequential streams of operations via the notion of *sequential consistency*. To quote his 1979 paper:

> A multiprocessor is *sequentially consistent* if the result of any execution is the same as if all of the processors were executed in some sequential order, and the operations of each individual processor occur in this sequence in the order specified by its program. [66, 25]

Sequential consistency often arises in describing what guarantees a parallel platform makes about instruction ordering, particularly in the presence of instruction level parallelism within a single processor core and in the presence of caches for multiple cores sharing a single memory. In both cases, the hardware protocols and logic are designed to guarantee sequential consistency.

The concept becomes important to a programmer when parallel code is written where an assumption is made that sequential consistency holds. When sequential consistency is important to guarantee correctness and determinism in the results of a parallel program, it is important that the programmer identify these regions as critical sections. The semantics of a critical section ensure that interleavings that would violate sequential consistency are avoided by controlling the execution of the section of code via locking or other synchronization mechanisms.

Caches and locality

While architectural topics are outside the scope of this text, there is an important architectural feature that must be recognized. In systems based on multicore or other tightly coupled shared memory processors that utilize caches to increase performance in the same manner as for traditional single processor designs, there are mechanisms within the memory subsystem that allow caches to behave as they are expected to. When performance is a consideration, in sequential programming the presence of caches requires the programmer to take the memory system into account when structuring operations such as loops to ensure that the code is written to exploit the spatial and temporal memory access locality that caches are built to optimize. A similar consideration must be made for caches when they exist in a multiprocessing environment in which coherence is provided to ensure sequential consistency at the instruction level.

Modern hardware provides a very transparent interface from the processor to the main memory that is actually quite complex. For efficiency reasons, in the 1960s computers began to use the concept of a *cache* to take advantage of what is known as *locality* in most programs. More often than not, a program that operates on one location in memory is very likely to operate on nearby locations in subsequent operations. This is known as *spatial locality*. Thus, the hardware brings whole regions of memory into a very low latency memory close to the processor, reducing the need to go out to slower main memory for most operations. Due to the small size of this cache, when later operations cause the hardware to bring new data close to the processor, older data that is no longer being operated on must be written back out to main memory so its space in the cache can be reused. Therefore we want code to execute that exploits spatial locality to execute soon after data has moved into the nearby, fast memory before it is forced to be written out to main memory and replaced. This proximity of operations in time is referred to as *temporal locality*.

The effect of caching is that the programmer often has no knowledge of whether or not they are operating on memory in the main store, some level of faster cache memory, or registers. The semantics of the hardware guarantee that data is in main memory when it is expected to be, but when this expectation is not assumed, the hardware is free to store data where it decides it is most efficient.

In multiprocessor systems though, this presents a problem. Say two processors are sharing memory. If one processor accesses a value, the values nearby are pulled into the nearby processor cache to reduce latency on subsequent memory operations. These subsequent operations will occur only on this cache memory, and updates on this memory may not be immediately visible in the original main store. If the second processor accesses a portion of memory that has been replicated into the cache of the first processor, it may not see updates that have occurred but have yet to be committed back to the main store. In this case, both processors may see different memory states, leading to incorrect data in one or more processes, and ultimately inconsistent and incorrect results in the programs. Memory and cache coherence is the term for hardware and software protocols that ensure that such inconsistencies do not occur.

We will discuss the protocols and performance issues that arise in maintaining memory consistency and coherence in chapter 8.

2.2.4   Thread safety

Most software is written in a modular form, as objects or simple functions, where sections of code are written without direct knowledge of the code that calls them. This is most apparent when looking at libraries, which are typically intended for reuse in a diverse set of applications that are unknown when the library is written. A consequence of this is that the writer of the library or function may be unaware of concurrency present in the program that will ultimately call the code. It is possible that the code will be called by multiple threads within a single program. When this occurs, there are potential consequences with respect to the correctness of the code being called. We refer to code that can be safely called by multiple threads in a concurrent program *thread safe*.

To illustrate what can cause thread safety issues we start by thinking about the concept of *side effects*. The following function contains no side effects — it will always produce the same result regardless of where it is called.

```
int square(int x) {
   int squared = x*x;
   return squared;
}
```

How do we say that this function has no side effects, and is not prone to effects elsewhere in the program? Upon examination we see that the function uses only memory that is local to itself. The parameter x is passed by value on the call stack, and the computation to be returned is stored in the local variable squared. Now, what if we rewrote the function to pass the parameter by reference?

```
int square(int *x) {
   int squared = (*x)*(*x);
   return squared;
}
```

Assuming that the compiler performs no optimizations and emits code that exactly matches what we wrote above, the function would dereference the pointer x twice. In a sequential program, this would rarely cause problems, as one and only one thread of control exists. While the function executes it can be certain that the value pointed to by x won't change unless the function changes it itself (which it does not). What happens when we introduce concurrency? One simple case that can appear in a single threaded program in modern operating systems is the effect of asynchronous I/O operations. What if x points at a location of memory corresponding to the contents of an I/O buffer for a network device? It is possible then that an incoming message could cause the value pointed at by x to change while the function executed. Similarly, a second case would be one where x points at memory shared by multiple threads of execution. While one thread executes the function, another thread may cause the value pointed at by x to change.

In both of these cases, the two places where x is dereferenced could cause two different values to be used in the computation leading to erroneous results. In this example, we see that a function can be thread unsafe if it makes assumptions about the state of memory outside of its direct control. Clearly a function to square a value like this is assuming that the value doesn't change while it executes.

Another type of safety issue that we frequently see in practice is due to side effects caused by the function itself. In the example above, side effects originated from outside the function. Consider the following example where side effects originate from within the function.

```
/* initialize as zero */
int lastValue = 0;

int foo(int x) {
   int retval = x + lastValue;
   lastValue = x;
   return retval;
}
```

This function takes an argument and adds it to the argument that was passed in to the last invocation of the function, or zero if the function had not been invoked yet. So, if we executed foo(4) followed by foo(5), we would see the return values 4 and 9 respectively. Each invocation of the function has the side effect of the *global* variable lastValue being updated. Assuming that foo() is the only code that modifies this variable, we would observe the function operating as expected in a sequential setting. When we introduce multithreading though, things can go wrong if multiple threads invoke the function. Why?

The use of a global for this purpose is to provide future invocations of the function some record of those that occurred previously without the caller explicitly passing this historical context in as an argument. Usually this context is meaningful, where within a single thread of execution repeated invocations of the function will require the context data to perform correctly. When multiple threads invoke this function, then the global variable will be written to by each of the threads, leading them to conflict with each other. If a thread executes the sequence above, foo(4) and foo(5), and another thread executes foo(20) in between, then the first thread will see 4 followed by 25 instead of 4 followed by 9.

We can fix this by forcing the caller to pass this context in. The consequence is that the caller must maintain the context itself and pass it explicitly to the function. The benefit is that the code with the context explicitly passed in becomes thread safe and no longer requires dangerous global data.

```
int safe_foo(int x, int lastValue) {
   int retval = x + lastValue;
   return retval;
}
```

This example is clearly contrived, but this precise problem crops up in practice. Many C programmers learn the hard way that the strtok() function for tokenizing strings is not thread safe. Why? The strtok() function must remember where it left off in the string being tokenized as it is called repeatedly. This is why the C standard library includes a second implementation, strtok_r() which includes an additional parameter representing the location where the last invocation left off. The suffix _r stands for *reentrant*. Reentrant functions are thread safe, while non-reentrant functions typically are not. Reentrancy simply refers to the ability of a function to correctly operate if another thread of execution invokes it while a previous thread has already started working with it.

2.3   Exercises

1. Consider the following program statements.

```
a = 4;
b = 6;
c = a+5;
d = b+7;
e = 7;
f = c*d;
g = f/e;
```

   For each statement S, specify the $IN(S)$ and $OUT(S)$ sets used to define Bernstein's conditions, and based on these conditions identify which statements can be executed in parallel.

2. A server is built that handles multiple incoming requests and uses a work queue to store them. A single worker thread pops tasks off of the queue, performs the corresponding action and responds to the client with the result. Does this server exhibit concurrency? Parallelism? Both?

Listing 2.4: A simple sequence of operations.

```
x = 4;
y = x+4;
x = y;
```

3. Consider a server similar to the one in the previous problem, but with two worker threads available to service requests that arrive in the queue. Does this server exhibit concurrency? Parallelism? Both?

4. Why is a multitasking operating system executing on a single core processor an example of a concurrent but not parallel system? Can there be aspects of the system (either hardware or software) that are actually parallel?

5. You find yourself needing to write a program to perform one operation for each element of a relatively small array. Would it be more efficient to perform this operation in parallel using independent processes or separate threads within a single process? Under what circumstance would it not matter significantly?

6. What properties of a block of code cause it to be necessary to execute atomically instead of as a non-atomic sequence of independent operations? Provide a simple example in pseudo-code.

7. The snippet of code in Listing 2.4 is executed in two threads of execution concurrently. Enumerate all possible sequential orderings of the operations across the threads.

8. Given the possible execution orderings from the previous exercise, what are the possible values of x and y at the end of execution on each thread?

9. Let A[] be an array of counters. Explain what can lead to A[i]++ being a non-atomic operation for a valid value of i, assuming no hardware assistance is provided (such as cache coherence) by the system.

Chapter 3

Concurrency Control

Objectives:

- Discuss problems that can arise due to poorly coordinated concurrent threads of execution.
- Identify causes of these problems and synchronization patterns that can be applied to control them.
- Introduce synchronization primitives used to implement these patterns.

We design and write programs to perform tasks, and one of the fundamental requirements of our programs is that they perform these tasks in a manner that we consider to be correct. For example, if we write a program that takes an input and produces some output, there are a set of possible outputs that are considered to be acceptable for a given input. If a program gives an output for a given input that is not acceptable, we would say that the program behaved incorrectly (most often, we say it has a bug).

Correctness is a subtle concept. In its most basic form, we think about correctness as a one-to-one mapping of input to output. Given an input, a single output is considered to be correct. This isn't always the case though. Consider a Web-based search in an intranet where pages are replicated across many servers. For a given query, any one of the instances of a page that matches it is considered to be valid. If the page mypage.html exists on server www-1 and server www-2, a correct response from the search program would be either http://www-1/mypage.html or http://www-2/mypage.html. The requirement of the search program to provide a correct response simply means that one of a family of correct responses is provided.

This type of correctness is often what we consider when dealing with concurrent programs. The behavior we see above in which multiple results can be produced for a given input is a result of nondeterminism. If multiple servers executed the search and the result to provide to the client was chosen from the first to respond, the value returned would vary depending on the load on the servers. In this example, this behavior is perfectly acceptable. On the other hand, what if we were interacting with a banking system executing a sequence of transfers, withdrawals, and deposits? Clearly from the perspective of both the bank and the account owner one and only one balance should remain at the end of the transactions. Nondeterminism in this case may be highly undesirable.

Concurrency control addresses this problem. What sorts of correctness issues may we see in a concurrent program, and what techniques exist for controlling these issues to write programs that behave as desired?

3.1 Correctness

Earlier in this text we discussed the basic concepts of concurrency, such as atomicity and the protection of critical sections of code via synchronization. In this section, we will introduce correctness issues that arise when assumptions such as atomicity do not hold and synchronization is not properly performed. These correctness issues are very important to understand. When designing programs that will execute concurrently, or libraries that will be used by potentially concurrent programs, one must consider the impact that this will have on the correctness of the code. The correctness issues here are those that must be considered whenever one is designing a concurrent program. They are also the core of the debugging process for parallel programs, as bugs will often have at their root one of these issues. Unlike sequential programs where many frequent bugs are due to errors in bounds, control flow conditions, or bad logic, bugs in concurrent programs arise from code that can appear to be correct in a sequential sense but leads to incorrect results due to interactions between concurrent or parallel threads of execution.

3.1.1 Race conditions

Race conditions are one of the most frequent errors that occur and generally appear unexpectedly, sometimes long after a program has been written and has functioned correctly for some time. What is a race condition? A *race condition* is a nondeterministic behavior of some portion of a concurrent program due to an un-enforced dependence between two or more threads of execution. If a thread $T1$ is written assuming that another thread $T2$ will have performed an operation (such as assigning a value to a shared variable) before $T1$ reaches some point where it requires the result of this operation, then the programmer must make this dependence explicit. If this dependence is not made explicit and enforced via some synchronization mechanism, then the execution of $T1$ may vary. This variation will depend on whether or not

Listing 3.1: A simple example of a race condition.

```
shared int counter = 10;

void thread1() {
    while (counter > 0) {
        counter = counter - 1;
        doSomeWork();
    }
}

void thread2() {
    while (counter > 0) {
        if (counter % 2 == 0) {
            printf("%d",counter);
        }
        doSomeOtherWork();
    }
}
```

an execution of the program occurred with an ordering that matches the dependence relationship between the threads. Race conditions arise when the programmer has assumed some ordering of operations between threads, yet has not coded it explicitly into the program.

For example, consider the code in Listing 3.1. Here we show two thread bodies that share a common counter variable. The first thread body (thread1) decrements the counter and does some local work until the counter reaches zero. The second thread body checks if the counter is even, printing a message if it is, and performs some other local work, repeating the process until the counter reaches zero. The output of this program is highly dependent on the manner by which the two threads execute. If, for example, the first thread body is extremely fast relative to the second, we may see zero or more printed outputs from the second thread. On the other hand, if their local work is similar in duration, we may see a few printed outputs from thread2. It is quite likely though that we will not see an output from the second thread corresponding to the counter values of 10, 8, 6, 4 and 2. There exists no synchronization between the threads that would force this to occur, so the best we can hope for is a lucky balance of work between the threads that would result in that output. The code as written does not enforce it.

Why is this a problem? Isn't it true that if the imbalance between the two processes is sufficiently large, we can assume that one will always reach the critical operation before the other? Of course not (otherwise, race conditions wouldn't be an issue!). As we said, race conditions often appear as unexpected

Listing 3.2: Code prone to deadlock.

```
void thread1() {
    lock(A);
    lock(B);
    work();
    unlock(A);
    unlock(B);
}

void thread2() {
    lock(B);
    lock(A);
    work();
    unlock(A);
    unlock(B);
}
```

behavior in code that had worked fine for long periods of time in the past. Often they arise after hardware upgrades that result in performance improvements of some portion of hardware or software. A new version of an operating system may improve the performance of an operation that was relied upon by the slower processes, leading them to catch up to the faster process. This is the source of the term "race condition." The processes are racing towards a critical operation, and the correctness of the program relies on them finishing this race in an assumed order.

How do we overcome race conditions? Often a race condition can be prevented by imposing synchronization on the critical operations that cause behavior changes depending on the order of arrival of the concurrent processes. This can be achieved by employing locks or forcing mutual exclusion around a critical region of code.

3.1.2 Deadlock

Deadlock is a common source of concurrent program failure, but unlike race conditions, is often easier to identify. This is because a program that deadlocks does not produce incorrect results — it simply gets stuck and ceases to progress in its execution. Consider the simple example in Listing 3.2. In this case, it is possible that `thread2` will acquire the lock on B before `thread1` has a chance to. This will result in A being locked by `thread1`, B being locked by `thread2`, and both becoming unable to proceed because they are each waiting on the other to give up a lock that the other holds.

Livelock

Livelock is, as the name should imply, related to deadlock in that the net effect is the program ceases to be able to proceed in its control flow. Instead of simply pausing forever waiting on locks that will never be released, a program experiencing livelock does experience a changing program counter, but a pathological interaction occurs between threads of execution resulting in a control flow loop that repeats forever. In some sense it is the computational equivalent to a real world experience most of us have encountered. Walking down a sidewalk towards someone, we encounter a situation where we must walk to the side to avoid a collision. Often both people choose to walk to the same side, resulting in an amusing (if uncomfortable) dance back and forth waiting for someone to break the symmetry and go to the other side so each person can resume their path down the sidewalk. Livelock is the case where this dance goes on forever, and instead of being stuck, we simply repeatedly perform the same operation over and over making no tangible progress.

Programmatically we can modify our deadlock example to become prone to an instance of livelock, as shown in Listing 3.3. Say that instead of blocking on the second lock acquisition, we first check to see if the lock is available, and if not, we unlock the lock we previously acquired and start over at the beginning.

An example of a real-world system where this behavior is possible is a transaction-based program, such as a database or one employing software transactional memory. The act of checking the state of a shared lock or variable, undoing some partial thread-local computation (in this case, the acquisition of the first lock), and restarting from the beginning, is very similar to the process by which transactions will undo and restart if a conflict is encountered with a separate thread of execution.

In the example shown here, the cause of potential livelock is a high degree of symmetry between the participants. The only way the code would proceed would be if there was a globally agreed upon order of lock acquisition and release (such as always acquiring a lock for A before attempting to lock B), or some degree of randomness in the time required to perform each step of the computation. For example, if the time between the lock acquisition and second lock test for `thread1` was sufficiently large (due, for example, to operating system preemption), it is entirely possible that the second thread would have time to release its locks. Of course, this would then require the first thread to acquire its second lock before the second thread attempts to on its retry.

This should make it clear to the reader that correctness problems of this form are very often highly dependent on *timing*. In a code more complex than the example here, it is very possible that the circumstances that can lead to livelock (or deadlock) are in fact very rare. This is often a very frustrating lesson learned by programmers of concurrent programs — a code that appears to work perfectly well, sometimes for a very long time, may suddenly hang for no apparent reason. Furthermore, the timing circumstances

Listing 3.3: Code prone to livelock.

```
void thread1() {
begin1:
    lock(A);
    if (locked(B)) {
        unlock(A);
        goto begin1;
    }
    lock(B);
    work();
    unlock(A);
    unlock(B);
}

void thread2() {
begin2:
    lock(B);
    if (locked(A)) {
        unlock(B);
        goto begin2;
    }
    lock(A);
    work();
    unlock(A);
    unlock(B);
}
```

Listing 3.4: Code prone to starvation.

```
void thread1() {
begin1:
    lock(A);
    lock(B);
    work();
    unlock(A);
    unlock(B);
}

void thread2() {
begin2:
    lock(B);
    if (locked(A)) {
        unlock(B);
        goto begin2;
    }
    lock(A);
    work();
    unlock(A);
    unlock(B);
}
```

may be very difficult to reproduce, making the process of debugging that much more infuriating. Fortunately, tools exist in the form of "thread checkers" or "thread debuggers" for assisting programmers in identifying these subtle bugs.

3.1.3 Liveness, starvation and fairness

Liveness, starvation and fairness are issues that arise in concurrent systems that have been studied for a long time, most often in the context of schedulers for shared resources. The issue in question here is whether or not a thread of execution can cease to proceed beyond a specific point *without* entering a state of deadlock or livelock. For example, consider a hybrid of the case studied in the previous sections where we make the threads deal with lock contention differently. We will make thread1 aggressive and greedy, waiting stubbornly for the second lock while holding the first, and immediately acquiring the second when it becomes available. On the other hand, we will make thread2 less aggressive, releasing its first lock if the second is unavailable in the hope that by allowing another thread to proceed it will be able to continue later.

If only two threads of execution were present, the net result of this structure will be a higher likelihood of thread1 proceeding first before thread2 when lock contention occurs. This may or may not have undesired side effects on the program (such as introducing a bias in a physical simulation), but from a control flow perspective, each thread executes as expected. The problem that arises is when many threads exist, all but one taking the aggressive approach. As the thread count increases, the likelihood of contention occurring for the locks goes up. Since the non-aggressive thread always backs off when contention occurs, it can become *starved* and never proceed into its critical region. A thread of execution is *starved* if it is not allowed to proceed beyond a point in its control flow due to other threads always acquiring exclusive resources necessary for the thread to proceed.

Liveness and fairness are related to starvation. Starvation is a situation that can occur, much like deadlock or livelock. Liveness and fairness on the other hand are not situations so much as *properties* of a system. A segment of code ensures *liveness* if it can guarantee that no thread can starve during its concurrent execution. A program ensures *fairness* for a set of threads if all threads are guaranteed to acquire exclusive access to primitive resources a reasonable percentage of the time. The precise probability of a thread gaining access to meet the criteria of fairness is often program dependent.

Starvation can occur quite often as a side effect of code that was written to avoid potential bad locking situations. Consider the case in listing 3.4. The second thread can starve because it releases a lock it acquires if it sees that the other lock that it requires has already been acquired by a different thread. While the code will not deadlock, an overaggressive instance of the first thread can cause the second to starve and never execute.

Code that is written to ensure fairness keeps this deadlock avoidance scheme, but attempts to give threads that backed off to allow other to proceed priority at later times. Often fairness is guaranteed by the underlying primitives or runtime system. As we will see soon in discussing semaphores, a mechanism for implementing synchronization and locking, the primitive itself can define an order in which threads that block are serviced. In the single lock case, a simple FIFO queue will guarantee a simplistic form of fairness, as no thread can force its way to the front of the queue before others that are already waiting. The situation becomes more complex to ensure fairness in cases where multiple locks are utilized as in listing 3.4, but it is still possible.

We will see these topics arise later when discussing language techniques that rely on transactional programming models such as software transactional memory (STM). In transaction-based systems such as STM or transactional databases, there is a concept of retrying a transaction in the case of conflicts. Starvation and fairness are important in these systems to prevent threads from encountering an endless sequence of conflicts that force repeated transaction retries.

3.1.4 Nondeterminism

Nondeterminism is a familiar term to computer scientists, and is often not considered to be a strictly undesirable feature. It is easy to conceive of programs where nondeterminism is actually an advantage from a performance and correctness perspective. It is often the case though that programmers make an assumption of determinism when they design and create their programs. Determinism simply implies that given a set of inputs, a program that is executed an arbitrary number of times on these inputs will always produce the same output by following the same sequence of operations to reach it. On the other hand, a nondeterministic program may consistently produce the same output, but it may do so by choosing different sequences of operations to execute. In programs where deterministic execution is required to ensure correct outputs, nondeterministic execution of the program can result in incorrect outputs being produced. As such, programmers must be careful when using nondeterminism in their programs to ensure that the output that is produced is correct under the various sequences of operations that can occur when executed.

An example of a situation where nondeterminism is perfectly acceptable, if not preferable, would be in a search engine that uses a farm of servers to execute a query. A client may submit a query that would be handed off to a set of servers that would then search the subset of data that they manage to find one or more elements (such as indexed Web pages) that match the query. Any matching element could be considered valid, so the first server to respond can provide a result that would satisfy the client. This would be desirable from the user standpoint, as the time-to-response is often important. Load variations on the servers, such as those seen when many clients may be submitting queries at the same time, will cause different servers to be the first to respond to a given query. Nondeterminism in this case would be considered to be a desirable feature of the system.

On the other hand, say we have a system that allows clients to submit requests to update a database and a set of servers that can service these requests that share a common database. An example would be a financial system, where update requests correspond to withdrawals and deposits. If two requests are submitted, one to withdraw X dollars from an account, and another to deposit Y dollars to the same account, the order that they occur could have significant impact on the results. Say a client has a bank account that contains 10 dollars, and submits a deposit of 20 dollars before submitting a withdrawal of 25 dollars. If two different servers pick up these requests to execute on the database, they must coordinate to ensure that the order in which the client submitted them is respected. If they do not, and execute in a nondeterministic fashion, then we have two possible outcomes. In one outcome, the operations execute in the order the client expected, and the final balance is 5 dollars. In the other outcome, the server managing the withdrawal request may execute it before the server managing the deposit executes. The result may be a rejection of the withdrawal or a penalty due to overdrawing the account. In this case, deterministic execution of the sequence of operations is very important — the set of servers must ensure that the order in which operations occur matches the order that they were submitted in.

Race conditions discussed earlier are a special instance of nondeterministic program behavior. When a race exists, the effect on the program output is dependent on which threads reached critical data access points first. The end product of such races is that the final output of the program will change as a result of the ordering that is seen at runtime. Without synchronization primitives put in place to enforce operation ordering that prevents races, the output of the program will be nondeterministic. This has long been possible in concurrent systems where multiple threads of execution were interleaved by the operating system or other runtime to give the appearance of parallelism through multiplexing shared resources. Unfortunately, in parallel systems the ordering is even more out of the control of the programmer and can result in timing differences or low-level hardware effects that can make execution times highly unpredictable. With the increasing presence of multiple processor cores in modern computers, the existence of truly parallel execution will make timing effects that were previously hidden by interleaving-based concurrency quite visible.

3.2   Techniques

So far, we've seen the concepts that arise in concurrent programs and the correctness issues that they carry with them. In this section, we will discuss techniques that can be used to address these concepts. The techniques are presented in the abstract. How one takes advantage of them is highly dependent on the programming tools at hand and the programming model being employed. The connection of these techniques to concrete language constructs will be made later in the text when we delve into constructs that are available in modern languages.

3.2.1   Synchronization

The fundamental activity that we require in concurrent programs to control interactions amongst concurrently executing units of execution is that of *synchronization*. Without synchronization, these units of execution are unable to interact in a controlled manner. The lack of this controlled interaction can result in correctness problems appearing in concurrent programs. Synchronization comes in many different forms.

Semaphores

Semaphores, described in detail shortly, are a simple method for synchronization. A semaphore is an object that is atomically modified using either increment or decrement operations, where one is defined as blocking when the semaphore has some special value, such as zero. As such, a semaphore implementation also includes a queue used to manage blocked operations to allow them to proceed in an ordered fashion as the state of the semaphore changes. A semaphore can be used to implement not only locks, but more general forms of synchronization. Say two processes are executing, and one needs to know at some points during its execution whether or not the other thread has reached a certain state (and possibly continued past it). If the threads share access to a semaphore that starts with a value corresponding to the fact that this specific state has not been reached yet, the thread requiring information about the state of the other can either check the value of the semaphore and continue appropriately, or block on the semaphore by calling the blocking operation. The thread that reaches the important state can simply execute the nonblocking operation to modify the semaphore, immediately either unblocking the other thread or setting the semaphore to the new value so the other thread may see it when it checks the semaphore next.

Mutual exclusion

Earlier we discussed the concept of a critical section which we wish to restrict access to. Mutual exclusion is the technique by which only one thread of control is allowed into this critical region of code. The mechanism by which this mutual exclusion is provided is typically via locking, which we discuss in the section 3.2.2. This is one form of synchronization. The concurrent threads of execution serialize their execution of this block of code should two or more attempt to enter it at the same time or while one is already in the region.

Rendezvous

Another method of synchronization is through a rendezvous point. Say two threads must wait for each other at some point in their execution, possibly to make an exchange of information necessary for both to proceed. This point in the program execution where this must occur is a rendezvous point. A rendezvous differs from mutual exclusion in that no thread is allowed to proceed once it reaches the rendezvous point in isolation, and progress only starts again once all participants reach the point. In mutual exclusion, the point at which the critical section is reached allows one thread of execution to enter, blocking others until it exits the section.

A non-computational example of a rendezvous is a team of athletes running a relay race. After each runner has run his or her portion of the race, he or she must pass a baton off to the next runner to allow him or her to start. The two runners must rendezvous at the point where the baton is passed. The rendezvous in this case represents the fact that the two runners must meet at the same place before they can continue — the runner with the baton cannot leave the track to rest until the runner picking it up is available, and the runner picking it up cannot start running his or her portion of the race until the previous runner reaches him or her.

If instead the runner with the baton simply had to place the baton on a table when he or she reached the end of his or her portion of the race and was able to start doing something else, while the subsequent runner could start anytime when the baton was available on the table even if the other runner had already left, we would have a different sort of synchronization that was *not* a rendezvous. That case would actually be closer to the type of synchronization we can implement with semaphores. In this case, there could be a period of time where the baton sat on a table with no runners around. Neither runner would be required to be at the same place at the same time to synchronize. This would yield a very different type of race, and obviously represents a strategy that would not yield a medal when competing with teams that appreciate the meaning of rendezvous synchronization!

The simplest form of a rendezvous is a synchronization barrier. A barrier simply indicates a point that the synchronizing participants must reach before being allowed to proceed. A basic barrier does not exchange information between threads. More sophisticated rendezvous mechanisms are available to not only synchronize control flow between threads, but to exchange information. We will see this employed by the Ada language in section 6.2.5.

3.2.2   Locks

Locks are a simple construct that most every reader should be familiar with, if not from computing, then from everyday life. At their heart, they simply are based on the notion of an entity (a lock) that can be acquired by one and only one thread at a time. In everyday life, we encounter locks on a daily basis. In a restaurant, a public restroom often has a single key. If someone has entered the restroom, the key is not available, and others who wish to enter must wait until the current user has returned the key to the counter.

Locks are intimately tied to the concept of atomicity. Clearly if the lock is intended to restrict access to some resource or region of code to a single thread, the action of acquiring or releasing the lock must be atomic, lest the exclusive access assumption be violated. A lock is used to protect critical sections and ensure mutual exclusion.

How locks are implemented varies, and is often dependent on features of hardware for performance reasons. In 1964, Dekker's algorithm was invented for the mutual exclusion problem showing that the only hardware guarantees necessary were that of atomic loads and stores. Modern hardware that supports operations such as "test and set" allow one to reduce the amount of code necessary to implement locks, resulting in platform specific performance

Listing 3.5: A busy-wait loop to acquire a lock.

```
successful = false;
while (successful == false) {
    do {
    } until (available(lock));
    successful = acquire(lock);
}
```

benefits.

One performance consideration that arises in lock implementation is how to deal with threads that are waiting for a lock to be released by a thread that has already acquired it. A simple method is to perform what is known as a *busy-wait*. We show an example of this method in Listing 3.5. In this listing, we do not employ a test-and-set operation to atomically check if the lock is available and acquire it. Instead, the lack of atomicity between the `available()` call and the `acquire()` call requires us to check whether or not the `acquire()` succeeds before proceeding to ensure that the lock is successfully acquired. It is conceivable that another thread may sneak in and acquire the lock in the short period of time between the end of the `do-while` loop and the `acquire` call, so this check is necessary. If the acquire fails, the `do-while` loop must begin again.

Atomicity aside, we see that there are performance implications regarding the separation of the availability check from the lock acquisition, leading to a potential for another thread to sneak in between the operations and acquire the lock. The thread waiting on the lock is repeatedly executing an operation to check the availability of the lock. For the entire period where the lock is not available, this loop executes, spending many CPU cycles checking the status of the lock. If the lock is held for a long period of time, a more efficient method to perform this operation is to somehow pause the thread waiting on the lock until it is available, leaving the CPU cycles available for other threads to take advantage of.

Implementing locks that do not require busy-waiting relies on the operating system or other layers to provide interrupts or other mechanisms that allow a thread to lie dormant until the lock acquisition occurs and the thread can be reactivated. Most systems provide this in the form of *blocking* calls. A blocking call will only return when it has achieved some task, such as acquiring a lock. The caller is frozen while the call executes, and the underlying method for acquiring the lock is hidden inside the call and can be optimized for the system implementing the lock. This optimized method is likely to have less performance impacts than explicit busy-waiting inside the caller.

Fairness and starvation

An issue that arises with lock acquisition in practice is that of fairness. As defined above, fairness is the concept that each process in a set of processes that is attempting to acquire a lock, or enter a critical section, will eventually get the chance to do so. Starvation is the consequence when fairness is not guaranteed. A thread that always loses to others in acquiring a lock or entering a region of code is considered starved, because it cannot make progress.

Different implementation methods can be taken to ensure fairness. For example, processes that are requesting a lock can be handed the lock in a first-in, first-out (FIFO) manner. Other disciplines can be chosen for problem specific performance reasons (such as those with varying levels of priority). Programmers who employ or implement locks must take fairness into consideration.

3.2.3 Semaphores

The concept of a semaphore was introduced by Dijkstra [27] as a flexible mechanism for implementing synchronization. A semaphore is a special variable that is modified atomically with operations to both modify and check the value of the variable, blocking the flow of control of a process if the value of the semaphore is in a specific state.

The primitive operations on semaphores are usually written $P(x)$ and $V(x)$. This can be a bit counterintuitive since one corresponds with incrementing and the other with decrementing the semaphore value. This notation is inherited from the original Dutch terms chosen by Dijkstra for operating on the semaphore [30, 31].

- **P = Prolagen:** To try and lower, derived from *Probeer te verlagen*, meaning "To try and decrease."

- **V = Verhogen:** To raise, to increase.

A semaphore can be represented as a simple integer to represent the state of the semaphore and an associated queue for keeping track of blocked threads. We start by considering the case of a *binary semaphore* where the values can only be 0 or 1. Operations on the semaphore s are then defined as follows.

- `initialize(s)`: s is initialized to the value 1.

- `P(s)`: If $s > 0$ then decrement s. Otherwise, the thread that invoked `P(s)` is blocked.

- `V(s)`: If another thread is blocked in a `P(s)` call, unblock the thread. Otherwise, increment s.

Listing 3.6: Protecting a critical section with semaphores.

```
void thread_body() {
  semaphore s;
  initialize(s);

  while (true) {
    work;
    P(s);
    critical_section_code;
    V(s);
    work;
  }
}
```

A simple snippet of pseudo-code for the body of a thread that uses a semaphore to protect a critical section is shown in Listing 3.6. As we can see, the first thread to encounter the semaphore passes through the P(s) operation unblocked, leaving the semaphore with a value of zero while it resides in the critical section. Any other thread that arrives at the P(s) operation blocks. When the thread that did not block completes the critical section and invokes the V(s) operation, one of the blocked threads that is waiting in the queue for *s* is removed from the queue and allowed to proceed. When the final thread exits the critical section, leaving no threads blocked on P(s), the value of the semaphore is returned to the value 1.

An alert reader should have realized that there is a subtle detail that has not been directly addressed related to fairness. How is the blocked thread chosen to proceed when V(s) is invoked if there are multiple threads blocked on the P(s) operation? A simple solution is to maintain a first-in, first-out ordering of the blocked threads such that they are unblocked in the order of their arrival. Other more sophisticated solutions can be taken, but from the perspective of the programmer it should not matter how the implementation ensures fairness.

Allowing for arbitrary positive values instead of simply 0 or 1 can lead to a more general purpose synchronization method sometimes called a *general semaphore*. A general semaphore (or *counting semaphore*) could be used to limit access to a small number of threads; for example, to limit the number of concurrent accesses to a Web page.

3.2.4 Monitors

Introduced by P. Brinch-Hansen [48] and refined by C.A.R. Hoare [54], monitors provide an abstraction originally intended for structuring operating systems. The key concept behind monitors is an encapsulation of data and procedures related to a component within a system that requires synchronized control when accessed by multiple callers. In the most basic sense, monitors provide an abstraction layer above basic semaphores, coupling the semaphore variables with the operations and data that the semaphore is used to protect. Monitors provide more than this though, through the availability of condition variables.

Monitors are an early instance of what object-oriented programmers refer to as *encapsulation*. The inventors of monitors recognized that complex codes that utilize semaphores to protect critical regions are difficult to maintain if the logic of managing the semaphore is spread throughout a program in an ad-hoc manner. Errors are easy to make if one accesses a portion of code or data that is intended to be protected by a critical region, but forgets to properly check and set the value of a semaphore. Global variables that require synchronization are a dangerous proposition and are highly error prone as software grows and changes.

As such, a monitor is a program structure that encapsulates the data that must be protected by a critical region, the synchronization primitives used to provide this protection, and the routines that operate on the protected data that must manipulate the synchronization primitives. Modern programmers who use languages such as Java with classes containing routines and data annotated with the **synchronized** keyword routinely implement monitor-like structures.

An important feature of monitors are *condition variables*. A condition variable is a mechanism within a monitor for synchronization of threads. Instead of lock-based protection, condition variables allow threads to signal each other to perform synchronization. A condition variable is associated with a monitor much like an object field, and represents a group of threads that wish to be informed of a state change in the monitor. Each condition variable provides two operations: **wait** and **notify** (sometimes called **signal**). The **wait** call causes a thread to enter the queue of threads that are waiting for the state change associated with the condition variable, and the **notify** call is used to send a signal to the threads waiting in the queue that the state change has occurred.

The important feature of condition variables is that when a thread executing within the monitor invokes **wait** on a condition variable, the thread yields access to the monitor and blocks.[1] This allows other threads to enter the monitor and perform the operations necessary for the state change to occur that the waiting thread requires. When one of these threads has reached a point where the waiting thread can be restarted, the **notify** call is invoked.

---

[1] Non-blocking variants of monitors exist, notably within the Mesa language. We only focus on the original blocking variant here. See [16] for an in-depth discussion of different types of monitors.

This allows the blocked thread to reenter the queue to reacquire the locks necessary to enter the monitor and pick up where it left off.

We see this pattern provided by some threading systems, such as Java-based threads. In Java, we can use the `synchronized` keyword to control access to methods or blocks of code. Within these synchronized regions, a thread may execute the `wait()` call, causing it to block and release the locks that protect the synchronized region. This allows other threads to acquire them and enter the synchronized region while the thread that called `wait()` remains blocked. When one of these threads executes the `notify()` call, the thread blocked on `wait()` will proceed to reacquire the lock and continue its execution. Let's take a look at an example of this in a simple object.

Producer/consumer with monitors

A classic problem used to demonstrate synchronization via monitors is the Producer/Consumer problem. We start with an object that represents a storage location (a "bin" in our code example) which acts as a monitor providing two synchronized routines — `set` and `get`. We want this bin to obey the following rules.

- The bin is empty initially.

- Any calls to `get` to access the contents of an empty bin will block until an element is placed into the bin by the `set` routine.

- When the bin becomes occupied, one and only one `get` call will be allowed to read the value. When the value is read, the bin becomes empty again.

- If the bin is occupied, any call to `set` will block to prevent the value in the bin from being overwritten.

So, how do we implement this? The important state change that we are using to synchronize producers and consumers is a flag indicating whether or not the bin is occupied. This is our condition variable associated with an instance of the `Bin` object. Consider the object shown in Listing 3.7. Consider the `set` routine first. Upon entry into the routine, the thread checks to see if the bin is full by looking at the value of the `isSet` variable. If this is true, the thread must yield the monitor to other threads to allow the value to be read by a call to `get`. Therefore the thread calls `wait` and blocks. When another thread invokes `notify` on the condition variable, the blocked thread wakes up and attempts to reacquire the locks protecting the monitor. As soon as this occurs, the loop iterates and the `isSet` variable is checked again. If it is now false, the routine continues and places its value into the bin, sets the `isSet` variable to true, and invokes `notify` to give threads blocked on `wait` a chance to proceed.

It is important that the `isSet` variable be checked after the `wait` call returns control to a thread. This is because more than one thread could be blocked on the condition variable in `wait`, and we have defined the bin as allowing one and only one reader to retrieve the value from it. The loop forces this check to occur when `wait` returns control to a thread to enforce this policy when there is competition between multiple threads for the `get` call.

The `get` routine is similar. We see a similar check of the state flag and entry into a `wait` loop, except this routine goes into the `wait` call when the bin is empty to allow a `set` call to provide it with a value. Once a thread successfully exits the wait loop, the value is read and the state variable is set to false. The `notify` call informs threads blocking on `wait` that they should attempt to proceed.

3.2.5 Transactions

The concept of a *transaction* comes from the field of database processing, and is intimately connected to the notion of atomicity introduced earlier. Recently, there have been interesting applications of the transaction concept from databases to the programming languages world, in which memory is treated as the equivalent of a database with the programs modifying it treated as clients. Techniques that use this abstraction in this sense are often referred to as implementing a *software transactional memory* (STM). It is important then to establish what a transaction is.

Often programs will need to update complex data structures (such as the mailing address example provided previously) in such a way that the entire data structure is updated in an atomic manner such that no reader will see a partially updated state. It often is the case that it is inefficient to acquire and release locks to protect the data structure and prevent all accesses to it by other threads, especially when it is highly unlikely that a write to critical data will occur by another thread during the execution of the complex set of operations updating the structure. Requiring mutual exclusion through strict lock acquisition and release around the entire sequence of operations can result in serialization of all accesses to the data, read or write, leading to a reduction in the effectiveness of concurrency within a program.

Instead, a transaction allows a process to enter into a region that it requires to be executed atomically, and work without locking by making all of its updates to a private store (often called a "log") that is not visible to other threads. All shared data that is read is recorded in the log, as are all writes made by the thread executing the atomic sequence of operations. When the sequence is finished executing, the thread can then compare the variables that it has read from in the shared store to those in its log to detect if they were modified by another thread during execution. This is referred to as *verification*. If no modifications are observed then the updates made by the transaction will be consistent and based on the state of memory shared by all threads. In this case, the thread can lock all data that it has modified, copy the modified values from its log to the shared store, and unlock them when it completes. This is known as the *commit* phase of the transaction. This sequence of locking, modifying, and unlocking in the commit phase can be made quite short relative to the time required to perform the atomic sequence itself. The result is less likelihood of serializing concurrent threads accessing a shared state. Furthermore, the use of a thread-private log during the sequence of operations prevents intermediate values from being visible to other threads.

One interesting feature of transactions arises when the verification phase detects that a shared memory location read from within the transaction has been modified by another thread. If this is observed, then any computations performed by the transaction using the out-of-date values that it had read before the other thread modified memory are now considered invalid. This makes the entire transaction invalid, and the results of the atomic sequence must be discarded. When this conflict is detected, a transaction can simply restart at the beginning, throwing out all data in the transaction log and beginning as if the previous execution of the atomic sequence had never occurred. The use of a private log makes this possible, as there is no possibility that the invalid intermediate computations could have been used by another thread. Potential liveness issues can occur if an unlucky thread happens to consistently be forced to restart a transaction. Implementors of transaction-based systems can put in place mechanisms to attempt to avoid or minimize instances of this situation.

Database developers have defined an acronym called *ACID* that states the requirements of a transaction. The acronym stands for:

- **Atomicity:** A sequence of operations that make up a transaction are considered atomic, meaning that either all of the operations complete or none of them do. No partial completion of the operations is allowed.

- **Consistency:** The state of the system that starts in a legal state before a transaction starts will remain in a legal state afterwards. This requirement is difficult to maintain in software transactional memory systems, as the notion of consistency is often application specific and not defined in the language itself. Consistency is easier to guarantee in software in database contexts where constraints defining legal state are part of the database definition.

- **Isolation:** Intermediate data during the transaction is not visible to other processes outside that which is performing the transaction. This is an important constraint to consider, as one may compute intermediate results during the course of a transaction that would be inappropriate for an outside process to access and use. Thread-private logs are used to implement isolation.

- **Durability:** Once a transaction has completed, the initiator of the transaction is guaranteed that the result of the transaction will persist.

In database terms, this means that the data has been committed and stored in a reliable storage repository. In software transactional memory, this may simply mean that the results of the transaction have been committed to the globally viewable state.

Obviously there is some interpretation as to how the ACID requirement fits with software transactional memory. For example, durability refers to the persistence of a transaction result in the presence of various system failures. This is not a reasonable assumption when the target state is volatile memory, as hardware or software problems that can result in program termination will obliterate the entire volatile memory of a program. In STM, durability is relaxed to simply mean that the transaction has completed and the result is guaranteed to be visible to all other processes that can access the memory acted upon by the transaction. Similarly, the requirement of consistency must be relaxed to account for the fact that, absent metadata that dictates constraints on the value of variables in a program, the requirement that a value be "legal" is a decision that the program itself must enforce. The STM runtime cannot enforce such legality constraints without additional information from the programmer.

3.3   Exercises

1. Say you have two multithreaded programs that hang due to a bug in their handling of critical regions. Looking at the process monitor in the operating system, we see that one of the programs consumes no CPU time while it is hung, while the other consumes a great deal. Which is a potential candidate for exhibiting deadlock, and which exhibits livelock?

2. Write two sequences of operations that access shared variables in which a race condition is present that can result in nondeterministic execution producing unpredictable results. The sequences should be relatively short, composed of three or four operations each. Make a note of where synchronization should be placed to prevent the race and make the sequences execute deterministically to produce consistent and predictable results.

3. The classic test problem in concurrency control is the dining philosophers problem. In this problem, you have N philosophers sharing dinner at a circular table. There is a plate of food in front of each philosopher. On either side of each plate, there is a single chop stick. Philosophers, of course, have excellent table manners so they need two chopsticks in order to eat. This means only a subset of the philosophers can eat

at any given time since a chopstick can be held by only one philosopher at a time. Design a synchronization protocol so these poor hungry philosophers can eat. Use the synchronization constructs described in this chapter. Discuss your solution in terms of deadlock, fairness and starvation.

4. OpenMP lacks any pair-wise high-level synchronization constructs since they violate the goal of maintaining consistent semantics with a serial execution of the OpenMP program. Using the flush construct (described in Appendix A) implement a semaphore to be used by a pair of OpenMP threads.

5. Using locks, design a FIFO queue that works with a single writer and a single reader.

6. Assume a system that supports atomic reads and writes of words in memory. Implement a FIFO queue with a single reader and a single writer but this time do this without using any locks or any other synchronization construct.

7. Demonstrate (or better yet, prove) that it is impossible to implement a correct FIFO queue with one writer and two readers without using locks when reads or writes are not atomic.

8. An increasingly important issue is to optimize a system so as to minimize the energy used during a computation. Explain how a program that uses transactional memory to handle synchronization might be more energy efficient than a program that uses fine grained locking.

9. In Listing 3.7, we see that the get() method invokes notify before returning the value stored in state. Is a race condition possible with threads that receive the signal due to the notify call with respect to the value stored in state? Explain why you do or do not believe a race exists.

Listing 3.7: A Java-based monitor implementing a synchronized storage bin for a producer/consumer problem.

```java
class Bin {
    private int state;
    private boolean isSet = false;

    public synchronized void set(int s) {
        while (isSet) {
            try {
                wait();
            } catch (InterruptedException e) {
                // handle exception
            }
        }

        isSet = true;
        state = s;
        notify();
    }

    public synchronized int get() {
        while (!isSet) {
            try {
                wait();
            } catch (InterruptedException e) {
                // handle exception
            }
        }

        isSet = false;
        notify();
        return state;
    }
}
```

Chapter 4

The State of the Art

Objectives:

- Motivate concurrency-aware languages by discussing current techniques based on message passing and explicitly controlled threads implemented as libraries.
- Discuss higher-level abstractions currently provided by some modern languages to aid in building concurrent programs.
- Discuss the limitations of traditional sequential languages with respect to automatic parallelization.

Concurrent programming is not a new topic, even though the advent of new processor designs has brought it to the attention of a wider audience in recent years. It is important for readers, in particular those with experience in concurrent programming or who have read other texts in the area, to understand the relation of the material in this book to the current state of the art in technology. This chapter discusses the popular techniques used at the current time and enumerates their shortcomings, especially those that high-level language constructs can overcome.

The current state of the art is largely dominated by three techniques: (1) threaded programming, (2) explicit message passing, and (3) asynchronous event processing. Threaded programming is a dominant model for realizing concurrency within application logic for general purpose programs such as games or business applications. Message passing on the other hand dominates the world of large scale parallel computing for scientific simulation where shared memory may not exist in the context of large processor-count hardware platforms. Message passing is also used to implement many distributed systems that are commonplace on the Internet. Event-based programming has long persisted in the user interface domain and has a strong presence in the network services area as well.

Given the maturity and persistence of these techniques, especially considering that many alternatives have been proposed over time, it is worth looking at them in detail to understand why they have persisted. While no concrete reason can be found to decide why they have consistently won out over alternatives, one can conjecture based on experience and observation. The most common reason that is given is that the low-level threading or message passing model is the most flexible for general usage that can be tuned for performance to application-specific needs. Event-based concurrency has persisted as it is simply a natural representation for the asynchronous actions that occur in user interactive programs or loosely coupled distributed programs.

A common line of reasoning (valid or not) found in sequential programming with respect to the use of C or other high-level languages that lie at the lower part of the language abstraction level hierarchy is the direct control by the programmer over performance related issues. Languages like C provide a minimal set of tools and rules, leaving programmers a great deal of freedom to tweak and tune the precise details of their algorithms to best suit the problems they are solving on the machines being used to solve them. Explicit threads and message passing provide a similar loosely constrained and fine grained view of the computational resources participating in a program. Unfortunately, with great power comes great responsibility. In the case of C-like languages, this responsibility requires very careful bookkeeping of memory and state to prevent bugs, security issues, and performance problems. Similarly, concurrent programmers who work at the low level must be equally attentive to details such as deadlock prevention, fairness, and determinism.

Our discussion of the state of the art will start with the lower-level, programmer controlled view of concurrency through explicit threads and message passing. We will then cover some higher-level abstractions that exist for programmers through database inspired transactions at the level of memory, encapsulation techniques such as the Actor model, and event-driven programming. A trend towards exploring higher-level alternatives has long been underway, and these techniques represent promising candidates in that direction.

4.1   Limitations of libraries

Before we dive into the current state of the art, we must discuss one of the fundamental reasons why many current techniques are limited from the languages point of view, even in the case of higher-level techniques. This is not an intrinsic issue with the models of concurrency that the current methods provide. Threads, message passing, and transactions are very viable methods for constructing concurrent programs. These *models* should not be called into question. The issue is with how they are *implemented* and how this implementation restricts automated language processing and compilation tools from both optimizing programs for performance and analyzing them for correctness.

Consider for a moment a program that uses functionality provided by some library, which we will call `libuseful.a`. This library can provide any number of capabilities, either for parallel computation, numerical processing, text processing, graphical interfaces — anything really. From the point of view of a programmer, the library provides a set of functions that are callable from their code. The library is utilized by making function calls that conform to the interface provided by the library. For example, we might see a block of code like:

```
x = 1*y*y;
j = 14.0;
z = UsefulFunction(x,42);
w = 88+j;
k = SomeOtherUsefulFunction(z,w);
```

What can we say about this block of code? First, in the absence of the actual source code of the library, the two functions that are called that are drawn from the library cannot be analyzed with respect to their function or, more importantly, any side effects that they may cause. So, most compilers treat them as fences — locations in the code where code reordering cannot cross. Code before a library call can be reordered such that the reordering takes place entirely before the call is made. Code cannot be reordered past the call. Similarly, we cannot assume the compiler will use any information within the library call to optimize the code on the calling side. In this example, it is possible that the first call to `UsefulFunction` will not actually use the first argument. The compiler cannot know this (without access to the source code for the function call), and cannot decide to eliminate or otherwise optimize the statement that assigns the value to x. So, the nature of libraries as "black-box" units of code means that optimizations are largely prohibited from taking any advantage of the contents of the library code itself. Of course, if we have the source code for the library it is entirely possible to take advantage of it at optimization and analysis time.

In the context of concurrent computation libraries (such as common message passing or threading libraries like MPI and POSIX threads), the limits are far worse. Not only are standard sequential code optimizations such as code motion and dead code elimination prohibited from working to their full capability, but optimizations across processors are impossible. What do we mean by this?

Say we have a program exchanging sequences of small messages between processors. In some cases, we might find that some of these smaller messages can be grouped together and sent at the same time to make more efficient usage of the connective infrastructure between the processors. Unfortunately, since message passing calls are implemented in the library and outside of the language definition that the compiler is able to perform analysis of, the compiler cannot provide any assistance for identifying potential optimizations such as this grouping. Optimizing code like this would be a manual process that the programmer would have to undertake.

Compilers treat all functions as equals — aside from potential optimizations on the language defined standard library, all library functions are black boxes that cannot be analyzed very deeply. So, a compiler cannot reason about properties of a parallel code based on library calls that implement the parallel operations. The only way a compiler can do so is if the parallel operations are promoted to being a part of the language, at which time the standard can dictate how a compiler can interpret and subsequently optimize the operations. When we discuss parallel languages we will see how operations provided by message passing libraries like MPI are represented by language primitives that the compiler *is* aware of, and can subsequently perform optimizations upon.

The vast majority of techniques for concurrency currently are add-on, library-like tools. While they are useful, their nature as third-party libraries limits the ability of automated tools to optimize their implementation. Moving these concepts *into* the language can make automated analysis, optimization and code generation possible. Extensions built on top of existing languages, such as those provided by OpenMP, are a compelling step in this direction in the context of C, C++ and Fortran. Before we dive into the language-based implementations of these techniques, we will first discuss their existing implementation in the form of libraries outside the specification of the language standards.

Finally, the issue of thread safety introduced in section 2.2.4 also arises when thinking about libraries in concurrent programs. It is important to consider how libraries interact with other code given their independent development. If they provide routines that cause side effects, it is very difficult to integrate them without modification into code that is based on threading, and subsequently sensitive to the correctness impacts of side effects. The lack of any facilities as part of the language standards for developers to address potential concurrency issues makes it difficult to anticipate and address them when developing a library. The language features that are generally lacking in a concurrency-unaware language include the ability to declare sequences of statements as atomic, or to define thread-private data to use instead of globals for maintaining the state that must persist between calls. Often the best a language provides for this purpose are keywords (such as `const` in C) that indicate that parameters to functions will not be modified within the function body.

4.2 Explicit techniques

Most techniques employed for concurrent programming currently require the programmer to explicitly define how the program code and data is decomposed across concurrently executing tasks, and establish the necessary synchronization and coordination logic to ensure that the program executes both correctly and efficiently. We will look at these explicit techniques for concurrent programming in this section.

4.2.1 Message passing

One of the oldest models of concurrent programming that traces its roots to C.A.R. Hoare is the model of *communicating sequential processes*, or CSP [55]. This work drew heavily upon earlier work by Dijkstra in the similarly named work on *cooperating sequential processes* [27]. The premise of Hoare's model is that processes are treated as independent entities, with their own state and program counter that is not accessible to other processes. Interactions required to coordinate execution or share data are explicitly stated via messages that are passed between the processes.

A simple way to think about this model is to consider a room full of people, each with paper and pencil available. Each person corresponds to a process. The task for the group to perform is decomposed into a set of smaller tasks that each person is assigned. To perform the global task, each person may need to communicate with other people to get information that the others have computed, and to share the results of their individual tasks. This communication can be accomplished by each person writing down some specific piece of information, either results that he or she has individually computed, or a request for results computed by someone else. To communicate this, he or she would simply pass a piece of paper (a "note") with the information or request on it to the appropriate other people. In turn, the recipients of the notes would read them and respond in turn, either by computing based on the new information received on the note, or writing a note in response with information that was requested.

The appeal of this model is that it fits well with how a group of people would solve a problem. A group of students working on a homework assignment will frequently say things to each other like "what answer did you get for problem 3b?," to which one may respond to the group or an individual, "my answer was 5.6." Solving the overall task (the homework problem set) is achieved by the group through a sequence of interactions where information is sought and answers provided via question and answer statements. The act of asking a peer for an answer, and the subsequent response with the answer, is an act of message passing.

Message passing is a fundamental abstraction that is conceptually easy to

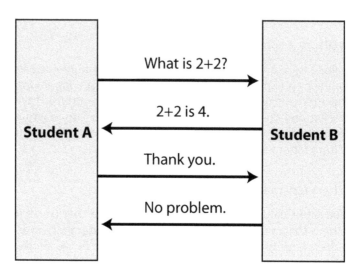

FIGURE 4.1: A sequence of messages between two students solving a homework problem.

work with, but has some tricky details that make efficient message passing programs challenging to write. For example, if one wants to compute the sum of a set of numbers distributed across a set of processors, efficient message exchange patterns based on combining partial results in a tree pattern can lead to logarithmic time complexity. Similarly, clever patterns exist for other operations such as broadcasting data to a set of processors from a single processor, having all processors broadcast a result to all others. It isn't feasible to assume all programmers using message passing will be intimately familiar with these tuned message passing techniques, so libraries began to appear that implemented these tuned and validated algorithms for off-the-shelf usage by programmers. Prior to these libraries, message passing primitives such as simple sockets were used to build distributed, network-based systems, but these provide very little abstraction to the programmer.

In the 1980s, a boom in computing technology based on the message passing programming model led to the creation of message passing libraries that provided optimized routines to programmers. In addition to taking advantage of general purpose algorithms with good asymptotic performance, these libraries could also provide platform-specific optimizations to use special features of different machines. While many libraries existed, two had the most lasting impact on the parallel computing community (primarily in the context of scientific computing). These were the Parallel Virtual Machine (PVM) [39] and the standardized Message Passing Interface (MPI) [3, 4]. These libraries

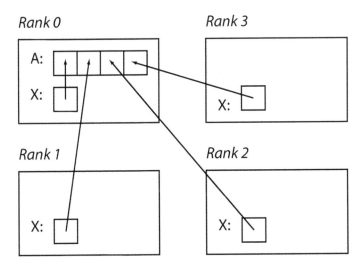

FIGURE 4.2: Accumulating an array of results from MPI processes onto a single master process. Rank 0 is commonly used for this sort of responsibility in MPI programs.

provided interfaces for C, C++ and Fortran programmers to build parallel message passing programs. During the 1990s, these two dominated the field, with some alternatives appearing and gaining smaller user communities (such as ARMCI [75]).

By the late 1990s and early 2000s, MPI dominated as the preferred message passing standard. MPI provides a few key abstractions that make building message passing programs easier. First, it provides point-to-point communications in both a synchronous and asynchronous model. This allows for programmers to optimize their programs to overlap communication and computation through the use of asynchronous messages and hide the impact of latency and communication time on their performance. Second, it provides a rich set of collective operations that can be implemented efficiently and take advantage of hardware assistance when present. Collective operations are a broad group of operations that involve more than pairs of processors and instead involve groups of processors. Finally, MPI allows for programmers to build and manipulate abstract groupings of processors and utilize logical communication channels that abstract above the physical channels that connect the processors together.

To illustrate message passing, let's say we have a set of P processors that all have computed a single integer value and want to send it back to a single control process that accumulates them all up into a single array. We can

Listing 4.1: A basic MPI-based C program corresponding to the array accumulation example in Figure 4.2. Calls to MPI functions are italicized.

```
int main(int argc, char **argv) {
  int myRank, numProc, src;
  double *values = NULL;
  double myValue;
  MPI_Status status;

  MPI_Init(&argc, &argv);
  MPI_Comm_rank(MPI_COMM_WORLD, &myRank);
  MPI_Comm_size(MPI_COMM_WORLD, &numProc);

  myValue = doSomeComputation(myRank);

  if (myRank == 0) {
    values = malloc(sizeof(double)*numProc);
    values[0] = myValue;
    for (src = 1; src < numProc; src++) {
      MPI_Recv(&values[src], 1, MPI_DOUBLE, src, 1,
               MPI_COMM_WORLD, &status);
    }
  } else {
    MPI_Send(&myValue, 1, MPI_DOUBLE, 0, 1,
             MPI_COMM_WORLD);
  }

  MPI_Finalize();
}
```

see this illustrated in Figure 4.2. In code, we would write this as shown in Listing 4.1.

This example has a few interesting features. We can see that there are functions called to initialize and finalize the MPI program, along with functions to allow a process to determine its place in the parallel program (its rank and the number of processes in the program, or "world"). The first interesting property, as pointed out earlier in section 4.1, is that MPI programs are based entirely on library calls that limit the degree to which automated compilation techniques can help the programmer with the concurrent aspects of their program. While the MPI calls all start with the string "MPI," there is nothing about them that is related to the language that the program is written in, so very little can be done to perform any specialized analysis on them.

Second, as the message passing is not part of the language, basic functions can be quite awkward. We must explicitly tell the MPI calls the location in memory for data participating in messaging operations such as sends or receives, the data type (in the form of an enumeration value instead of information from the language type system), and the number of elements contained within the message. This additional type information that the programmer explicitly provides is necessary for the calls to determine how to pack and unpack the data sent in a message. This doesn't have to be the case though. As we will see in Erlang (a message passing declarative language), sending a message can be as simple as a single statement:

```
ReceivingId ! SomeData
```

This single line sends the value contained in `SomeData` to the process identified by `ReceivingId`. The typing and size information about the message data to transfer is handled by the language and runtime, as are the details about the actual location in memory where the data is stored. It is safe to assume that the increased interest in Erlang and similar high-level languages in recent times with the arrival of multicore is due to the fact that it is significantly cleaner and conciser than previous techniques (such as MPI) for building concurrent programs based on a message passing model. Many of the tasks that MPI makes explicit to the programmer are implicit in higher-level languages.

Outside the scientific community, other message passing techniques have been created for building distributed systems. The use of basic TCP/IP sockets is quite commonplace, with abstraction layers provided to insulate the programmer from the details of the actual socket manipulation and management. These layers typically assist in initiating connections, dealing with failures, and marshaling data from the program into the form necessary for transmission, and back out again on the other side. Most modern languages provide this sort of abstraction layer through standard libraries or add-on packages. Unlike the more specialized MPI-like libraries, socket abstraction libraries typically do not provide routines for a rich set of parallel collective operations. Programmers using these libraries must implement them manually, or use third-party libraries that provide efficient group communication mechanisms.

Remote method invocation

A technique that shares some properties with message passing is that of remote method invocation (RMI) or remote procedure calls (RPC). An RMI or RPC system allows programs to cause operations to occur in remote processes through familiar function call syntax. For example, in Listing 4.1, we saw that a set of processes all send a single value to a master process to store in an array at the location corresponding to the rank of the process providing the data. This requires writing code such that an explicit send and receive pair is invoked. Instead, we can hide the messaging under a layer giving the appearance of regular language-level function calls. The master process (rank zero in the MPI program) can provide a function to set the value of a specific element of the array, and make this function available through an RMI or RPC layer.

```
void setArrayElement(int index, int value) {
    values[index] = value;
}
```

The processes providing the data could use the remote invocation layer to place their values into the corresponding location on the master. All that would be required of them would be to use the RMI or RPC system to acquire a handle that is used to communicate with the master process, and then use this handle to invoke functions that are syntactically identical to their local counterparts.

```
handle = AttachToMaster();
handle.setArrayElement(myRank, myValue);
```

The key difference between explicit message passing and the RMI or RPC layer is that they hide the details related to data marshaling, movement and actual function invocation on the side that actually executes the function. Arguments that are passed from the caller to the callee are essentially messages, and the method being invoked provides the context for what the messages are to be used for. One can implement an RMI/RPC-like system with messages alone, but would have to manually write the necessary logic for marshaling data, interpreting the intended method to invoke when a message arrives, and making the actual function call. RPC and RMI systems simply package this into a nice, clean abstraction layer that can be quite tightly integrated with a language and hidden from the programmer. The RMI system provided by Java or CORBA are examples of this, and are widely used in the software industry today.

The difference between RPC and RMI is largely in the context in which the invoked function actually executes. In RMI, as the name indicates, methods associated with object instances are available to be executed remotely. On the other hand, RPC (the older of the methods, originating in the days before object-orientation was ubiquitous) allows for procedures to be invoked that are not associated with specific objects. From the perspective of messages traversing the connective fabric between the caller and callee, the two are not conceptually different.

One significant shortcoming of this approach to concurrent programming is that remote invocations do not behave exactly like local invocations. In particular, there are failure modes that can occur when invoking a function remotely that are not possible in a local context. For example, if a connection is lost or the process providing the routine to invoke crashes, the function call on the caller side may fail to occur. RMI and RPC systems provide mechanisms to programmatically deal with failure modes like this, so these cases can be worked around. The big issue is that the use of the same syntactic form (function invocations) for two different types of invocations with different behaviors can be confusing and problematic when writing code. Code that is written based on function calls within the same process can experience very different error conditions when some of these functions are replaced with remote invocations to other, potentially distributed, processes.

Readers who are interested in RMI and RPC are encouraged to refer to texts on distributed systems (such as [24]) or specific implementations to learn more about these issues. We will not discuss them in detail here.

4.2.2 Explicitly controlled threads

This section has a curious name that some readers may find intriguing. Often programmers who work with threads simply refer to them as just that, threads. Why call out "explicitly controlled threads" as the programming technique in such a verbose manner? The reason is that the complicating factor with the technique is not the threads themselves, but the fact that programmers explicitly handle their creation, synchronization and coordination. Like instruction scheduling of machine code, the technique itself it not flawed. It is simply the fact that the programmer should not be working at this level in all but the most exceptional cases. Machine code is emitted by compilers in all but the most specialized cases, not by the programmer.

Evidence of the potential for problems that comes from working with low-level thread details can be seen in the evolution of the Java threading system. The `stop()`, `suspend()`, and `resume()` methods of the `Thread` class were deprecated due to their unsafe nature. Alternatives to their use suggested in the Java documentation are intended to avoid problematic situations that could occur when these methods were used incorrectly. Threads were not the source of the problem, but instead the difficulty of using them correctly and safely.

When we introduced the definitions of threads and processes earlier (section 2.1.1), we discussed that threads exist within processes and share memory. Due to the correctness issues that can arise with shared access to a single store (such as non-determinism), programmers must be careful to manage access to this shared state to preserve the overall correctness and determinism of their programs. When programmers must manually manage locks and critical sections, we call this "explicit control." The language and compiler do nothing to assist the programmer to ensure correctness. It all must be performed (and subsequently debugged and tuned) manually.

The most prevalent threading implementations in use are based on operating system level threads (Windows threads, Linux threads) or user level threads (POSIX threads, Java threads), which themselves are likely to be implemented on top of those provided by the OS. In these thread libraries, the programmer is able to define blocks of code that are to execute within a process as independent threads. Java threads are a special case in this set of thread technologies employed in practice because the compiler can assist the programmer in managing correctness issues. The availability of the synchronized keyword allows the programmer to specify parts of the program that must be treated as special in a threaded context to prevent problems. The compiler and Java Virtual Machine are then responsible for the management of the underlying locking mechanisms.

Say a program uses shared memory to represent a set of values that must be updated atomically. The classic example of transferring money between two bank accounts is useful to consider here. In this example, we have two bank accounts represented as floating point variables holding their respective balances. Operations such as transfers require atomicity such that no thread can observe the value of either account while one is being debited and the other credited. The sequence of withdrawals and deposits must appear to be a single operation. To make this happen, we restrict access to the balances of the processes by requiring a lock be acquired for both the source and destination accounts before the transfer may occur. Only when a thread has acquired both locks may the transfer actually take place.

As discussed earlier, it is easy to make mistakes in implementing this sequence of lock acquisitions, leading most likely to an instance of deadlock. This can occur in a situation where a lock for one account is held by one thread, and the other held by a different thread, causing both to block pending the acquisition of the lock currently held by the other. When programmers explicitly manage threads, they must take care to both identify potential places where conflicts such as this may occur, and write their code in a way that prevents erroneous situations like deadlock. The goal of a higher level language is to use abstraction to decouple the operations on shared data from the mechanisms put in place to ensure that they operate in a deterministic, correct manner.

4.3   Higher-level techniques

While message passing and threading are the dominant methods for implementing concurrent programs, there are higher-level abstractions that are also in use that remove some of the burden from the programmer that lower-level techniques impose. These techniques are built on top of threading or message passing abstractions, and are intended to address some of the issues that we have discussed up to this point.

4.3.1 Transactional memory

The difficulty and danger in programming for concurrency is primarily based on issues that arise when coordinating threads with respect to their access to a shared memory. Threads can be safe so long as they reference only thread-local data. The danger occurs when one thread is trying to read from data that another thread is currently modifying. This can lead to violations of atomicity and subsequent non-determinism in the execution of a thread-based algorithm. Traditionally, locks have been used to limit (serialize) access to a common data store. As introduced in section 3.2.5, transactions were invented as an alternative to explicit locking and mutual exclusion. Transactions allow speculative execution of atomic regions of code to occur, keeping the intermediate results safely isolated from other threads until the transaction completes. At completion time, the transaction system verifies that no conflicts occurred with other threads during the execution of the transaction and atomically commits the updates from the log to the main shared memory. If a conflict did occur, the transaction system can retry the transaction body as though the conflicting run had never occurred.

Recently, the notion of software transactional memory (STM) has been invented to reduce the complexity associated with using locks to control access to data [86]. STM systems take the concept originally intended for databases and make it available within general purpose programming languages to provide concurrency control for managing the shared memory in concurrent programs. Few existing languages actually provide an `atomic` block keyword, so STM capabilities are instead provided through library calls or compiler macros that translate into library calls. On the other hand, a quick Web search turns up many current projects that are using STM either transparently in a language implementation, hidden from the programmer, or are experimenting with the addition of keywords to the language to support transactions. For example, Intel has released a prototype compiler for C and C++ that extends the base language with keywords for defining atomic regions of code. Similarly, the Haskell community has demonstrated an elegant application of monads for implementing an STM system [50] in the purely functional Haskell language. The Clojure language also uses STM for supporting concurrent programs [45].

Software transactional memory works well when the probability for memory contention is rare. If this is true, the likelihood of a conflict is low, and more often than not the sequence of operations that occurs in a transaction will be valid and will be committed to memory without any issues. If thread contention over memory is a common event, then STM mechanisms are probably not appropriate. In this instance, the use of software locks and the extra cost in complexity may be worth the effort.

4.3.2   Event-driven programs

Every program cannot be assumed to be based on threads that have known synchronization patterns or knowledge of other threads that they may interact with during their lifetimes. A prime example of this is the common graphical user interface (GUI), in which programs execute autonomously until either the user or another thread requires the program to interact with it. A common model for this mode of execution is based on *events*. The model is based on activities occurring called events, such as a mouse click or an update being available for an on screen graphical display. The sources of these events are called *providers*, while the programs written to react to them are known as *consumers*. The key to the success of this model in many instances is the ease by which asynchrony can be introduced in terms of the event generation and consumption. Threads that produce events do not necessarily have to block until the events are consumed, and consumers do not have to block to wait for potential events to arrive. While it is *possible* to implement an event-based program this way, the event model does not require it.

A simple way to write an asynchronous event-based program is to implement threads that consume events as either single threaded programs that use system primitives such as `select()` to periodically poll an event queue for incoming events, or use a second thread that exists solely for the purpose of receiving events and making them available to the consumer. The second case is more conducive to parallel execution, as the polling-based methods are essentially a form of cooperative threading where the control thread periodically yields to the event handler explicitly. The thread that handles events that arrive can invoke event-specific code that modifies the state of the computational thread to influence its execution. Clearly this requires careful synchronization, but the abstraction has some nice advantages.

Consider the graphical display example where a GUI window may be processing user input or performing computations on a buffer that it is displaying. If a second program is the source of the buffer data that is displayed, it can update the GUI by generating an event that represents a buffer update (possibly encapsulating the new data within an update event object). The handler on the display program that listens for buffer updates can, upon receipt of an update event, lock the buffer, replace the data with the new contents, and unlock it. The locking would ensure that the display program would never be performing computations on incomplete or inconsistent buffer data, and the locking mechanisms to protect the data would allow the program to behave well in the presence of asynchronous, unpredictable events.

Obviously this example is over-simplified and prone to problems. An over-aggressive program that generates many events faster than the display program can handle them could cause the display to cease to be usable, and complex computations that use the buffer data without locking coarse regions of code could be disrupted if a buffer update occurs without the event handler taking into account the state of the computation. It should suffice though to illustrate the concept of an event-based system, in which event producers are coupled to consumers, with handlers that process events as they are generated allowed to operate asynchronously from the program that is ultimately using the information that is carried with the event.

4.3.3 The Actor model

Another popular model of concurrent programming that is based on message passing can be found in real systems. This abstraction is known as the Actor model [51, 5]. We will defer much of the coverage of it until later when we discuss a popular concurrent language based on the model, Erlang, but given its use in current systems, it is fair to say that it is part of the state-of-the-art. The Actor model is an abstraction above message passing that simplifies the details related to message handling on the sending and receiving sides of an interaction.

In many ways it is similar in spirit to the concept of a monitor, which simply provides a layer of encapsulation around concurrency control primitives and the state that they protect. An actor encapsulates message buffers and their sending and receiving operations inside of an abstract structure referred to as a *mailbox*. Threads, or actors, in the strict Actor model are implemented as independent threads of execution that do not share memory with others. These actors each provide a set of mailboxes into which other threads can asynchronously deposit messages. The absence of a shared state eliminates many potential concurrency control problems from occurring.

The mailbox provides the thread with a set of accessor routines with either blocking or non-blocking semantics. Threads wishing to communicate with others simply issue messages that deliver data to the target mailbox. The actor abstraction hides the underlying message passing from the programmer, allowing it to use a true message passing implementation or use a more optimized method for a given target platform (such as using shared memory). Like monitors that ease the difficulty of implementing correct lock-based code by hiding the locking within an object, actors hide the difficulty of handling both endpoints of message-like interactions. Not only does this ease the responsibility put on the programmer to ensure correctness and performance, but when the model is brought to the language level (as in Erlang), it allows compilers and runtime systems to further optimize how the abstraction is actually implemented.

4.4   The limits of explicit control

Why are we concerned with what we have termed "high-level constructs for concurrency" when it appears that we already have a mature set of tools that are widespread and well known? The reason is quite simple. At some point the complexity of programs reaches a level where automation is necessary to assist programmers in achieving good performance while maintaining a code base that is, for lack of a better term, manageable.

In sequential programs it would be highly undesirable to maintain programs at the level of assembly language simply for performance reasons when the size of programs reaches the complexity of a video game or desktop application. Instead, we maintain the program in a high-level language that is concise, manageable (via modular structures, separate compilation units, etc.), and in a form that allows an automated tool to perform the translation to high performance machine code. The equivalent of this is not possible for parallel code as currently programmed.

Consider the common and popular libraries such as MPI for message passing and POSIX threads or Windows threads for threading. To a compiler, these libraries are indistinguishable from calls to any other library. Absolutely no knowledge of the concurrent execution model in which the program will run is available to the compiler. As such, no optimizations can be performed that take into account the concurrent nature of the program and performance characteristics of the target platform related to concurrent execution. Similarly, no analysis is performed to reveal potential synchronization errors between threads that can be detected through static analysis at compilation time. As programs scale and the interactions between concurrently executing threads become more complex (especially when concurrent libraries are embedded into concurrent programs), ensuring both correctness and performance becomes very difficult to manage by hand. Adding a new thread to an existing multi-threaded program requires an intimate understanding of how this thread may adversely impact those that are already present, and many potential sources of error or performance degradation arise.

The introduction of language level features for concurrency into programming languages is precisely to address this issue. Like compilers for sequential languages facilitated the world of software that we currently see around us, compilers for parallel programming will be necessary to see a similar proliferation of software that is truly exploiting concurrency. To do so, the abstraction of concurrency must be a first-class part of the language that is accessible for analysis by the compiler. The black-box library approach is unable to function in this capacity for compilation. That said, the library approach is showing no indication that it will be leaving any time soon, and for good reason — in many cases, it is sufficient to get the job done.

In between libraries and fully-concurrent languages are extensions of exist-

Listing 4.2: A potential aliasing situation.

```
// declare variables
int a,*x,*y;

// ...
// some sequence of code setting x and y
// ...

// now, code that works with x and y
a = *x;
*y = 42;
a = *x+1;
```

ing sequential languages such as OpenMP. In this case, code in an existing sequential language (C, C++, and Fortran) is annotated with information related to concurrency that the compiler is free to use. This allows compilers that are aware of these extensions to perform some level of analysis to automatically target a parallel platform. We must make clear though that these annotations do not turn existing sequential languages into true parallel languages. The reason for this is that there are features of sequential languages that are fundamentally prohibitive to compilation techniques for optimizing concurrency in a large code.

4.4.1 Pointers and aliasing

One of the easiest examples to see of this conflict between existing sequential languages and parallel compilation is that of pointer aliasing. In fact, the imperative nature of most popular languages in which memory is directly modified causes aliasing to be a significant obstacle to automatic parallelization of codes written in these languages. A *pointer alias* is a pointer that refers to memory that another pointer also refers to.

The basic issue behind pointer aliasing is that compilers must be careful not to perform optimizations that involve a set of pointers that would change the meaning of the program should they point at the same location in memory. A very simple example is shown in Listing 4.2.

Before the lines of code that read from *x and write to *y are executed, some sequence of operations occurred that determined the values of the pointers x and y. In the first line, the value pointed to by x is read and stored in a local variable. In the third line, this value is accessed again and a scalar value is added to it. On the second line though, a value of 42 is written to the address pointed to by y. A common optimization that is useful for parallel code generation is to reorder code such that fine-grained parallelism can be revealed through simple operations such as a topological sort of the control and data flow graph.

Unfortunately, it is easy for a compiler to find itself in a situation where it is unable to determine if x and y are aliases for the same location or not. They could have been assigned in a subroutine call that the compiler cannot analyze and recognize the presence of the aliasing, or their values may be input-dependent. If they were aliases, the second and third lines could not be swapped as that would violate the required order of operations. Aliasing complicates the task of identifying dependencies within source code. As discussed earlier, the identification of dependencies is a critical stage in identifying parts of a program that can be executed concurrently.

The heavy use of pointers in code written in C and C++ makes the aliasing problem a source of difficulty when attempting to parallelize code in these languages. Mechanisms do exist to annotate variables with attributes (such as the `restrict` qualifier in C99) that inform the compiler of properties such as the lack of aliasing for a pointer.

4.5   Concluding remarks

As we have seen, there are a few reasons why many current languages and concurrency techniques leave something to be desired when it comes to concurrent programming. We have library-based techniques that prohibit analysis and optimization of the concurrent structure of a program, language annotations that provide concurrent primitives but are crippled by inherent features of the underlying sequential language, and convenient techniques for managing and manipulating memory that severely limit automated parallelization.

For the remainder of this book, we will discuss and demonstrate a few higher-level language abstractions and languages that implement them to demonstrate that these limitations of the common languages and concurrency techniques can be overcome. In the next chapter we will examine high-level language constructs from popular languages and discuss the implications they have for concurrency. We will follow this with a discussion of the development and evolution of concurrency in programming languages and a look at high-level constructs for concurrency available in some modern languages.

4.6 Exercises

1. You are given a message passing library that allows your code to allocate buffers and hand pointers to them to the library so that data will be placed into them when messages are received. What concurrency control issues may arise between your code and the message passing library with respect to these buffers?

2. Describe a situation where a thread may starve in a concurrent system that utilizes software transactional memory (STM). How would you use a different concurrency control mechanism than STM to avoid this starvation?

3. Write a simple message passing program in MPI to pass an integer token in a ring around a set of processes, with each process adding its ID (or "rank") to the value of the token. The originator of the token (typically at rank 0) should print the value of the token when it makes a single pass around the ring. What is the value?

4. Write a threaded program using either POSIX threads, Java Threads, or Win32 threads to compute the minimum and maximum value of a large shared array initialized with random numbers. One and only one thread should print the results when all threads complete and combine their results into the global result.

5. Using Java RMI, write a thread class that implements a simple instance of the Actor model. The class should provide methods that can be called by remote processes to deposit messages in the mailbox of the actor. The main body of the thread should check the mailbox and invoke the appropriate code locally to process the messages and perform the necessary computations.

Chapter 5

High-Level Language Constructs

Objectives:

- Examine familiar constructs from sequential languages and their role in concurrent programs.
- Understand the impact of side-effects on concurrent programs.
- Discuss the cognitive aspects of programming language constructs to establish a framework for thinking about the design and impact of concurrency constructs discussed in the remainder of the text.

Programming languages originated with the need for an easier and more productive method for programming early computer systems. The goal was to use abstraction to hide both the target hardware and internal structural complexity of a program. Hiding hardware allowed for programs to move between different computer platforms without significant modifications by relying on the compiler developer to understand the mapping of language abstractions onto specific targets. Hiding software complexity allowed for programs to become better organized and more complex by hiding the tedium of the low-level implementation of common patterns like loops and branches. Abstractions provided by languages included:

- Primitive operations and keywords that may map to a sequence of simpler assembly language operations.

- Looping operations that abstract above the primitive sequence of loop tests and branching operations for iteration and escaping the loop.

- Defining reusable sequences of statements as callable, parameterized subroutines.

As computer scientists appreciate, abstraction is a powerful tool for dealing with complexity in designing algorithms, tools, and languages. From the languages perspective, abstraction separates the expression of an operation by the programmer from its ultimate implementation in machine code. The three examples above illustrate this, as each allows the programmer to express a complex operation in simple terms and defers the difficulty of correctly and efficiently translating it to machine code to automated tools-compilers.

For example, consider the following code:

```
int main() {
  int i,j=1;

  for (i=0;i<33;i=i+7) {
    j = j * i;
  }
}
```

This rather simple code that utilizes the common `for`-loop control flow abstraction is translated when compiled without optimizations to the following x86 assembly by the GNU C compiler:

```
          .text
.globl _main
_main:
          pushl   %ebp
          movl    %esp, %ebp
          subl    $24, %esp
          movl    $1, -12(%ebp)
          movl    $0, -16(%ebp)
          jmp     L2
  L3:
          movl    -12(%ebp), %eax
          imull   -16(%ebp), %eax
          movl    %eax, -12(%ebp)
          leal    -16(%ebp), %eax
          addl    $7, (%eax)
  L2:
          cmpl    $32, -16(%ebp)
          jle     L3
          leave
          ret
          .subsections_via_symbols
```

The notion of "higher-level" constructs corresponds to raising the level of abstraction that programmers work in when writing code. Familiar layers of abstraction are shown in Figure 5.1. At the lowest level a programmer writes machine code in binary, which in modern times is extremely rare and often unnecessary. This binary code is either translated within the processor to a simpler microcode or passed along directly to the underlying functional units and logic gates that actually implement the instructions. Assembly language lies at a higher level of abstraction, in which binary code is expressed in mnemonic forms that give easily remembered names to the raw binary strings fed into the processor. This translation from mnemonic form to binary is performed by an assembler. These mnemonics are architecture specific, so they serve to make programming easier for a specific platform but have no impact on portability.

Sequential programming languages such as C abstract this even higher, where expressions and operations in the language are mapped by a compiler to the architecture-specific assembly and binary code for the machine. The development of languages has led to many useful abstractions. Type systems were created to provide a mechanism by which the compiler could assign properties to data elements which could then be used to ensure that operations on the data and groups of data elements made sense within their context. Organizational abstractions such as structured programming and object orientation provided powerful methods to structure programs and data to both deal with the complexity of large programs and, in the case of object oriented programming, represent meaningful entities that are program-specific and do not exist in the type system of the language. Developments such as parametric polymorphism built upon these techniques to further reduce the amount of code necessary to implement complex algorithms and data structures. The following toy example in C++ is a perfect example of this sort of reduction of code by high-level abstractions through the use of templates (which are similar to generics in other languages, such as Java).

```
template<class T>
class simple {
public:
  T a,b;
  simple(T x, T y) { a = x; b = y; }
  T add() { return a+b; }
  T sub() { return a-b; }
  T equal() { return (a==b); }
};

int main() {
  simple<int> s(4,5);
  simple<float> t(4.0f,5.0f);
  simple<char> r('a','b');
  simple<double> v(3.14,2.72);
}
```

This code translates to well over 300 lines of assembly, growing each time a new instance of the simple class is instantiated with a new type. In this case, we see four instantiations made inside the body of main(). The compiler is emitting the necessary code to support the different types that the class is instantiated with. The power of the abstraction of C++ templates is that the

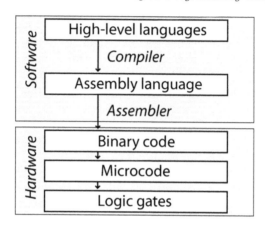

FIGURE 5.1:   Abstraction layers of programming.

programmer does not have to write code for each possible type instantiation of the templated class. The compiler is responsible for this.

In this chapter, we will look at high-level language constructs, what motivates their design and the impact concurrency has on existing constructs and the design of new ones.

5.1   Common high-level constructs

We will briefly review important high-level constructs present in existing programming languages and pay careful attention to the impact that concurrency has on them. This impact can be both positive and negative. On the positive side, some constructs open the door to identifying and exploiting parallelism. On the other hand, the meaning of some constructs changes in the presence of concurrency in a way that can impact the overall correctness of an algorithm implemented using them. We will not cover the full range of high-level constructs that have been invented over the history of programming language development, and instead touch upon those most commonly implemented in languages and used by programmers.

Readers interested in a deeper history of high-level programming language design should refer to programming languages texts such as Sebesta [85] or Scott [84]. Detailed treatments of the development and evolution of languages that had a significant influence on language evolution can be found in the proceedings of the ACM History of Programming Languages (HOPL) conference series.

5.1.1   Expressions

The first programming languages were focused on the predominant problem facing computer users in the early days of computing — defining complex mathematical operations for automatic computation. For example, consider the following contrived formula in four variables:

$$f(a,b,c,d) = \sum_{i=1}^{50} (a+i)^3 * \left(i*b + \frac{c}{d}\right)$$

From a programmatic point of view, this is actually quite easy to program in modern languages. We would simply have a loop that iterates a variable i from 1 through 50, and a single statement body like:

```
sum += pow(a+i,3.0) * (i * b + (c/d));
```

We take for granted that the language provides statements that allow us to express the mathematical operations with arithmetic operators in the proper order and grouping via parentheses. Similarly, the languages provide mathematical primitives such as the exponentiation function, or others such as basic trigonometric functions and logarithms.

Before these early languages, one would have to decompose this expression into the primitive operations that it is composed of:

```
tmp = c/d;
tmp2 = i*b;
tmp = tmp + tmp2;
tmp2 = a+i;
tmp3 = a+i;
tmp2 = tmp2 * tmp2;
tmp2 = tmp2 * tmp3;
tmp = tmp * tmp2;
sum = sum + tmp;
```

An abstraction as simple as basic expressions reduced the amount of code necessary to express complex mathematical operations, and furthermore, deferred the management of intermediate results (indicated above in the manually coded sequence as tmp variables) to the compiler. Similarly, the order of computations and precedence of operations was defined by the language and enforced by the compiler in the code that was generated. If we want to add to this the additional code required to implement the loop for the summation operation, and the required pre-loop variable allocation and initialization, the sequence of primitive machine operations will get quite large.

Trivial as it may seem to start with expressions, even this most fundamental of language abstractions has consequences in the context of concurrent programming. Expressions are not necessarily atomic. In the example above, we see that the single expression is actually implemented as a sequence of simpler subexpressions that are generated by the compiler. Clearly if atomicity of an expression is important, then we must take care to protect complex expressions just as carefully as we would a large critical region. If one writes an expression such as

```
x = a[1] + (a[0] * (b[1] / a[1]))
```

which is ultimately compiled to a sequence of operations, what happens if the subexpression for b[1] / a[1] executes, and a[1] is modified by another task before the second reference to a[1] executes as part of the addition operation? Recalling the notion of atomicity, we must consider how expressions are treated by a compiler of a parallel language in the presence of concurrent tasks with a shared state. Most sequential languages are specified independently of potential use in threaded environments, and do not specify that atomicity of expressions be enforced, especially if the expression contains references to memory or operations that cause side effects. In the absence of threading as part of the language, programmers must enforce atomicity where necessary using their chosen threading library.

Side effects

In the example above we see that interleaving of the operations that make up a complex statement with those in another stream of execution can lead to inconsistent results due to non-deterministic execution of multiple instruction streams. What is the ultimate source of this?

In our simple example, we have an expression that references data in memory to perform its computation. The issue that we are faced with is if the state of this memory changes during the execution of the sequence of operations that implement the expression. A change in memory is what is known as a *side effect* of a computation. Pure computational operations do just that — compute. Operations with side effects (*impure* operations) do more than compute in that they change the state of the world around them. Side effects can range from modifying the state of memory explicitly to performing I/O operations. The ability to write code with side effects is provided by nearly all languages, especially those with imperative features. As such, these languages leave open the possibility of writing code that is inherently prone to errors that result from side effects.

The impact of side effects is felt by programmers differently depending on what type of language they are working with. The important distinction between languages is whether or not the programmer has explicit control over how the program actually runs on the computer, or defers this decision making to the compiler. We distinguish between languages based on this factor by classifying a language as either *imperative* or *declarative*. Modern languages have been blurring the line between the two classes for many years, but it is still useful to think about. Languages tend to fall more on one side than the other, such as the ML family of functional languages, which encourage purely functional, declarative programming, but allow for side effects when necessary. On the other hand, C is a very imperative language, but provides the `const` keyword to enforce a declarative-like single assignment discipline for specified variables. Java provides a similar facility via the `final` keyword.

Given a task that we wish to solve, when using a language that adopts the imperative model we tell the computer precisely how to go about performing the sequence of computations to solve the task. This includes how data might be arranged in memory, what order sequences of statements occur, and so on. On the other hand, languages that are based on the declarative model require that we tell the computer what it must do to solve the problem in a more abstract sense and defer the details of memory and data management, instruction scheduling, and other lower-level tasks to the compiler or runtime support system.

The difference between the two styles is largely in the abstraction level at which the programmer works. In imperative programs, the developer has fine control over many details about how the program actually executes on the computer, such as how data structures are built, how arrays are organized, and what sequence of operations will be performed. In concurrent programs, this means that the programmer has direct control over how data is distributed and shared between parallel processing elements, and more importantly, how the concurrently executing processes interact with each other such that correctness is preserved and high performance is achieved.

Functional, logic, and dataflow programming languages are forms of declarative languages. These languages often prohibit (or, strongly discourage) side effects because such effects complicate reasoning about them in a formal setting. Those that do allow for side effects typically attempt to restrict what types of effects are allowed to occur and carefully control how they interact with the pure effect-free core language. The lack of side effects makes automatic parallelization significantly easier for the compiler.

5.1.2 Control flow primitives

As referred to above, early languages also introduced control flow primitives to provide abstractions that are present in most algorithms. These include:

- Looping constructs based on conditional tests such as `while` loops.

- Looping constructs based on conditional tests with iterate variables, such as `for` loops.

- Conditional branching in the form of `if-then-else` branches.

- Conditional branching on multiple conditions in the form of `switch` statements or pattern matching.

Control flow primitives are necessary because nearly all algorithms are not simple sequences of operations all executed in order. Very often there exists a subsequence of operations that must be executed only when some condition holds, and an alternative sequence is provided when the condition does not hold. Similarly, repetition of a sequence of operations is usually present when the algorithm operates on a set of values, or requires a repeated operation on a single value to arrive at the final answer.

In the context of concurrency, there is potential for parallelism in some of these abstractions. Consider a loop that performs some operation on each element of an array (such as adding one to each element). On a traditional CPU, we would expect this loop to be implemented as a sequence of assembly instructions for the body of the loop preceded by a test for the loop termination condition, and terminated with a simple counter increment and jump instruction back to the termination test. On the other hand, in the presence of vector (or SIMD[1]) hardware, we could see the entire loop absorbed into a single vector operation, or a loop of vector operations. In a third (although not final) case, we may allow compiler optimizations such as loop unrolling to eliminate the loop entirely, restructuring the code to eliminate unnecessary jumps and conditional tests.

The key to the success of abstractions such as arithmetic expressions and the `for`-loop is that users can assume that the compiler will ensure that the resulting binary code will behave as they expect. The semantics of the `for`-loop construct will be preserved regardless of the manner by which it operates. This is the intended consequence of abstraction. The implementation is separate and independent of the construct, but the semantics of the construct are guaranteed by the tool that generates the implementation through a language specification that the compiler conforms to. The language defines evaluation ordering and operator precedences for expressions, and defines the control flow behavior associated with loops. Similar language specifications exist for all high-level constructs, and provide a contract between the programmer and the compiler that guarantees an expected behavior that the compiler will attempt to implement as efficiently as possible within the constraints of the specification.

5.1.3   Abstract types and data structures

Quite rapidly after the introduction of abstractions for expressions and control flow specification, language developers introduced the notion of types associated with data in programs. At the fundamental hardware level, particularly in terms of the representation in memory, all data is represented equally

---

[1]We define SIMD and related terms later in section 6.1.8.

as sequences of binary bits. The fact that a programmer may intend one sequence of bits to represent a 64-bit single precision floating point number versus a set of 8-bit ASCII characters is unknown to the memory hardware. Somehow programmers must ensure that when they intend to use a sequence of bits from memory in a specific way, the appropriate operations are executed in the processor to implement it. When the data in memory represents more complex types (such as vectors or strings) than are natively supported by the processor, additional effort is required to ensure that operations on the data are implemented correctly so as to preserve the data abstraction intended by the programmer.

We frequently write programs using these data structures and types that are not supported natively on hardware. Processors have no concept of a string, so languages provide abstractions that allow programmers to work on strings while translating the operations into low-level manipulations of byte sequences. Similarly, record or structure data types exist to allow programmers to treat heterogeneous groupings of low-level types as units at the language level. For example, complex numbers can be used by C++ programmers through the use of classes and operator overloading, and by Fortran programmers via the intrinsic `COMPLEX` data type. In all of these cases, the languages provide a mechanism to define structures that are meaningful to the programmer, and automatically translate them back and forth from representations on the hardware.

Each of these representations are built out of more rudimentary pieces. A complex number is nothing more than a pair of real numbers. A string is a sequence of characters, possibly augmented with a length value. Again, these familiar abstractions have consequences in the context of concurrency. Recall the example we showed above in which a single statement was decomposed into a long sequence of operations. In that case, whether or not a language guarantees that this sequence will execute as an atomic unit has significant impacts on correctness in the presence of concurrent execution. A similar problem exists in the context of complex data types.

Let's consider for a moment the simple example of a complex number. A complex number can be treated as a two-dimensional vector with two components, r and i. If we multiply a complex number by a scalar value S, then the result will be a complex number with components Sr and Si. When computing the scalar product with a vector, we again have the same problem of atomicity that arises due to the two separate multiplications that make up the overall operation. Like the sequence of operations that was discussed earlier, in which a value in the expression may change during execution of the expression, when working with complex data types we may find ourselves in a situation where the variable has not been completely updated before another task attempts to access it.

The concept of monitors discussed in section 3.2.4 is a mechanism to deal with this. The data structure encapsulates the complex set of data elements that must be updated in a controlled manner. It protects them by forcing accesses to occur through a set of functions that hide the locking mechanisms used to guarantee atomicity of important operations. Many languages designed for concurrent computing provided monitors and monitor-like data structures as part of their definition. Objects that use the `synchronized` keyword in Java are an example of the use of a language feature inspired by the monitor concept.

5.2   Using and evaluating language constructs

Now that we've briefly thought about the constructs that make up languages we are familiar with and some of the issues that arise with relation to them in concurrent contexts, let us step back for a moment and think about what makes a useful language construct. If we are to evaluate new constructs intended specifically for the purpose of expressing concurrency in our programs, then we should establish a framework for thinking about and evaluating them. In this section, we introduce a novel approach to this question by looking at recent work in the cognitive aspects of programming.

In writing a program, you start with a problem specification and map it onto a sequence of instructions that execute on a computer system. We can think of this as an information process in which information from a problem domain is mapped onto the highly constrained domain defined by the architecture of the computer.

It is the job of the programmer to manage this information process. Unfortunately, models natural to human reasoning are poorly supported by the models inherent to the architecture of a computer system. We close this cognitive gap with high-level language constructs. These provide high-level abstractions that make it easier for programmers to carry out their appointed tasks.

For example, consider the need for a *critical region* in concurrent programming. This is a very common construct in parallel algorithms; a block of code that must be executed by each thread, but only one thread at a time can safely execute the code. Consider how this construct is represented with two commonly used parallel programming notations — POSIX threads and OpenMP.

The POSIX threads API (pthreads) is a low-level programming notation. To support a critical region, the programmer must write code with explicit locks as shown in Listing 5.1. The pthreads programmer must think about not only the sequence of operations that make up the critical region, but he or she must also define and initialize lock variables and explicitly acquire and release them.

On the other hand, OpenMP is a higher-level notation that augments the

Listing 5.1: Explicit locking to protect a critical section in C with POSIX threads.

```
int sharedA , sharedB;
pthread_mutex_t swapMutex;

void ThreadBody(void *threadid) {
  int tmp;

  pthread_mutex_lock (&swapMutex);
  tmp = sharedA;
  sharedA = sharedB + 4;
  sharedB = sharedA - 2;
  pthread_mutex_unlock (&swapMutex);

  pthread_exit(NULL);
}

int main(int argc, char **argv) {
  /* ... */
  pthread_mutex_init(&swapMutex, NULL);

  for (t=0; t<NUM_THREADS; t++) {
    rc = pthread_create(&threads[t], NULL,
                        ThreadBody, (void *)t);
    if (rc) { /* ... */ }
  }

  for (t=0; t<NUM_THREADS; t++)
    pthread_join( threads[t], &status);

  pthread_mutex_destroy(&swapMutex);
  pthread_exit(NULL);
}
```

Listing 5.2: An equivalent program using OpenMP to implement the same critical section as Listing 5.1.

```
int sharedA , sharedB ;

int main ( int argc , char **argv ) {
    int t ;

#pragma omp parallel private(t) default(shared)
    {
        /* ... */
#pragma omp critical
        {
            t = sharedA ;
            sharedA = sharedB + 4;
            sharedB = t - 2;
        }
        /* ... */
    }
}
```

base languages (C in this case) with compiler directives to label regions of the program as parallel and sequences of code within those as critical regions. One can easily argue that removing the explicit locking makes the code easier to write (there is less of it), and easier to read (the critical region is clearly defined, not inferred from the presence of locks). The OpenMP critical region is also inherently more reliable because the programmer cannot make a mistake in acquiring and releasing the locks — the OpenMP implementation hides that. We can see this demonstrated in Listing 5.2.

We can go further. OpenMP only supports a restricted subset of algorithms that map onto its fork-join, SPMD or task queue programming models. The compiler directives augment the base language with concurrency constructs, but they do not penetrate deep into the language itself and support more complex optimizations by an interpreter or a compiler. Another approach is to build the synchronization primitives into the language itself. This was done in Java. In Java, one puts the code corresponding to a critical region within a synchronized block of code. Locks and compiler directives are avoided and the integration into the language makes the program easier to write but also easier to read.

Tradeoffs between high-level and low-level language constructs are common in parallel programming. For example, in section 4.2.1, we compared the explicit send and receive pairing for MPI programs, with their detailed arguments, to the more concise exclamation mark notation used by Erlang.

In each of these cases, we can see that the cognitive load placed on the programmer varies widely when accomplishing similar tasks in different notations. But these overly simple comparisons are dangerous. By focusing on one factor (how easy is it for the programmer to write the code) we potentially miss a host of issues that become critical later in the programming process. If we are going to balance the strengths and weaknesses of different notations, we need to establish a conceptual framework that we can use to drive our comparisons.

Sebesta [85] addressed this problem by suggesting three primary criteria to consider when evaluating programming constructs.

1. *Readability*: This is a metric of how well a program can be read and understood by a programmer. The balance in language design sought by readability is a set of constructs that a human can read, and that a computer can interpret and emit an efficient implementation for.

2. *Writability*: Writability is intimately tied to readability, but implies that the language allows programmers to write their domain-specific programs using abstractions that facilitate readability.

3. *Reliability*: Reliability means that a program behaves as specified in all situations. Languages provide facilities to assist with this through type systems and exception and error handling mechanisms.

We can use these criteria, for example, when considering the issues that arise when working with constructs at a lower level of abstraction. Machine code and assembly language are hardly readable. Even if one is proficient in reading mnemonic codes and understanding the primitive operations, it requires significant effort to unravel the actual algorithm that the code was intended to implement. Similarly, taking an algorithm and decomposing it into primitive operations and manually lowering the level of abstraction can be quite tedious. Language aids to assist with these problems are measured by their ability to enhance readability and writability.

Reliability is also an interesting way to look at what a language provides. Languages themselves often don't make programs reliable — in any language, poor programming or design can yield incorrect programs that are unreliable. The issue here is that the language assists the programmer by providing assurances that the fundamental building blocks from which a program is built will execute reliably. Type systems are an example of this. In a strongly typed language, if one has a variable representing a string and another representing a floating point number, the language will emit an error or warning if the two are combined in an arithmetic operation like multiplication. Without the type system providing this check on the programmer's code, unusual things may happen at runtime if data of incompatible types are combined in unspecified ways. So, the type system aids in reliability by checking that the programmer cannot write code that doesn't make any sense from a type safety standpoint.

Similarly, exceptions in the language can be used to ensure that programmers provide code to deal with issues that may arise during execution. In older, lower-level languages like C, error conditions required the programmer to check return values or global variables for error conditions and to act on them accordingly. If the programmer didn't check the value of a return value or status parameter, or globals like `errno`, the compiler won't complain beyond a possible warning about return values being discarded. So, C doesn't help programmers identify places where they really should be providing code to handle errors or other exceptional cases. On the other hand, if a routine may cause an exception to occur in languages like Java, the compiler will notify the programmer that they must deal with the exception with a `try-catch` block, or pass the exception on to a parent routine. Exceptions in the language give Java the ability to assist the programmer with reliability issues that other languages like C do not.

Sebesta's criteria are useful for high-level comparisons of different programming notations. But to study the nuances presented by different high-level constructs, we need a more fine-grained framework. For this we turn to the cognitive dimensions.

5.2.1   Cognitive dimensions

The central problem is to understand how well the information systems defined by different high-level languages interact with a programmer. This issue is as much about psychology as it is about computer science. An influential researcher in the psychology of programming community is Thomas Green. After years of working on cognitive models of programming, he shifted gears to consider notations and the information they encode [42]:

> The way forward is not to make strong, simple claims about how cognitive processes work. The way forward is to study the details of how notations convey information.

He developed a framework [44] to facilitate discussions about information systems called the "Cognitive Dimensions Framework." This framework defines distinct features that characterize the constructs that make up a programming language. We will use a subset of Green's dimensions as we analyze different high-level language constructs.

Viscosity

Viscosity in a physical system describes how the system resists change due to an applied force. Similarly, programming languages with high viscosity resist changes by programmers. Green defines a viscous system as one in which many actions must be performed by a programmer to accomplish one goal, such as renaming a method in an object and having to chase down all usages of the method throughout the code base. The problem of viscosity in programming notations is well understood. Modern integrated development environments (IDEs) recognize the high viscosity of modern programming languages and often include tools to automate common refactoring techniques.

Visibility

Visibility describes how easily a system allows components to be viewed. Higher levels of encapsulation reduce visibility. In some cases, this is undesirable when the encapsulation hides information that the programmer needs access to. On the other hand, a system with low visibility can in fact be beneficial since it can prohibit the programmer from gaining access to low-level primitives that, given access to them, can lead to correctness problems. The example of monitors discussed in section 3.2.4 is a prime example of this — the act of reducing visibility of the data protected by the monitor reduces the likelihood of concurrency-related correctness problems.

Premature commitment

Programmers encounter premature commitment when the programming language forces them to make decisions before sufficient supporting information is available. Premature commitment problems arise when (a) the notation contains many internal dependencies, (b) the software development environment constrains the order of doing things, and (c) the order is inappropriate. For example, OpenMP programmers working on a non-uniform memory architecture must initialize their data such that the memory pages are close to the processors that will do the computation. But often, the way work will be scheduled onto processors is not known until much later in the program than the definition and subsequent initialization of data structures. Programmers deal with this by putting in a quick-and-dirty initialization and then, after the key computations are scheduled, they go back and redo the initialization to support the needed schedule.

Hidden dependencies

A hidden dependency is a connection between two entities that is not apparent in the text of the program; i.e., the dependence is hidden from the programmer. A classic source of hidden dependences arises when multiple inheritance is used in C++. A programmer may depend on a feature of a base class without any knowledge of the details inside the base class. This is a deliberate feature of C++ and is a key value to working with class hierarchies. But, unknown to the programmer, a feature in a base class may be changed or present an unexpected artifact in a programs execution. This can result in unexpected and erroneous execution and since the source of the change is hidden from the programmer, fixing these problems can be extremely difficult.

Role-expressiveness

This dimension measures how readily the purpose of an entity can be inferred. For example, compare the code in Listings 5.1 and 5.2. In the POSIX threads case, one must identify the locks, their acquisition and release, and the code and data encapsulated within the region where the lock is held to infer what the critical section is that they are protecting. On the other hand, the OpenMP case makes the critical section explicit by prefacing the code block with the critical directive. We would reason that these two systems would score differently when measured for role-expressiveness.

Error proneness

This dimension is fairly clear by its name. An error prone system invites mistakes and does little to protect from them. Languages that expose pointers explicitly present a good example of an error prone system. For example, when working with pointers, it is easy to introduce bugs where an invalid pointer is dereferenced, and these problems are notoriously difficult to debug. Information notations can be designed to reduce cognitive slips that lead to errors. They can also deal with the problem by providing an effective feedback system to help the programmer find errors once introduced. For example, a mistake in C yields the less than informative *segmentation fault* or *bus error*. On the other hand, Java will throw a *null pointer exception* when a mistake is made, giving the programmer the opportunity to write code to gracefully deal with it, or at least identify the location in the code where the exception originated. At the other extreme are languages where direct referencing of memory is prohibited entirely, and these errors simply do not exist.

Abstraction

Abstraction mechanisms build upon lower-level notations to create higher level notations suited to a specific program or activity. Abstractions make key features of a system explicit while hiding less essential features. Computer scientists learn early on that abstraction, often in the form of data structures or abstract data types, yields programs that are easier to write in the face of so much complexity. When talking about the abstraction cognitive dimension, we distinguish between *abstraction rich* and *abstraction poor* notations. For example, C++ with its ability to customize class hierarchies to suit the needs of a particular problem is abstraction rich compared to C. On the other hand, programming languages that force excessive amounts of abstraction can be difficult to learn; in essence, the effort to manage required levels of abstraction presents a so-called *abstraction barrier*. For example, Threading Building Blocks (TBB) [82] which manage concurrency through full-featured templated libraries, require the programmer to establish a host of base classes to do even simple concurrent computations. Hence, TBB is an abstraction rich notation but with a high abstraction barrier.

5.2.2 Working with the cognitive dimensions

The cognitive dimensions are tightly coupled. An information system is pulled in different directions as it attempts to balance the impact of different cognitive dimensions. For example, increasing the ability for a programmer to build abstractions to support a specific programming problem increases the potential for hidden dependencies in a program. The two dimensions pull the design of a programming language in different directions, and this is okay. What we have accomplished with the dimensions is to give names to these trends so we can systematically identify, discuss, and optimize the tradeoffs we must make as we consider which high-level language constructs to use in a programming language.

The cognitive dimensions define a vocabulary that can be used when discussing the effectiveness of different high-level constructs during the programming process. In [43] they evaluate these dimensions relative to a set of very specific programmer activities. These include:

- *Transcription*: Converting, copying, or translating something from one form to another. For example, turning a set of mathematical equations into a program is a transcription process.

- *Incrementation*: Adding to something to an existing program, such as adding an object to a set.

- *Modification*: Modifying an existing entity within a program; for example, changing the computation within the body of a function.

- *Exploration*: Rapid prototyping to explore a new design idea or test ideas about the understanding of a system.

The cognitive dimensions interact with these activities in different ways.

Transcription is fundamentally associated with creating a new program from scratch, either from the mathematics of a problem or a program specification. Hence viscosity and visibility, dimensions that pertain to working with existing code, are not key dimensions when considering transcription. On the other hand, transcription is greatly complicated by languages with high abstraction barriers and premature commitment.

Incrementation is all about modifying existing code. Therefore, the activity of incrementation benefits from programming languages that have low viscosity and minimal amounts of premature commitment. An abstraction rich environment can be helpful during incrementation if the incremental changes to the program do not cut across abstraction classes and hence run into significant abstraction barriers.

Programmers spend more time maintaining code than writing it in the first place. Maintenance is tightly linked to the activity of modification. Making modifications to a program is greatly complicated when the programming language displays high viscosity and if it forces the programmer to expend

significant effort dealing with high abstraction barriers. To make modifications to a program, the programmer must find where the changes need to be
made (visibility) and track down all the places in the program that might
be impacted (hidden dependencies). Finally, the already difficult process of
modifying the code is all the more complicated when the information system
forces one to make premature commitments.

Exploration is fundamentally tied to rapid prototyping; either explicitly
when constructing experimental programs or mentally when reading a program and trying to understand its meaning. The exploration activity is greatly
complicated by anything that limits the flexibility of making changes to a
program (viscosity, premature commitment, and abstraction barriers). A rich
abstraction system can be helpful as can a system with high visibility.

Role expressiveness is of vital importance when learning an information system. If a programmer can easily infer the meaning of a construct based on the
other constructs in the system, learning is much easier. Notations that empha-
size good role expressiveness add value for every activity associated with the
programming process as they make programs more understandable and hence
easier to locate where changes need to be made to support incrementation or
modification.

Error proneness cuts across all programming activities as well. An error
prone notation makes a programmer more likely to introduce bugs in the original code and to introduce errors as modifications or incremental enhancements
are added.

5.3   Implications of concurrency

When creating new languages, designers have sought constructs that were
orthogonal, implying that they did not interact in unexpected or difficult
to understand ways. For example, a simple arithmetic operation such as
addition does not affect the operands — adding two constants behaves the
same as adding two variables or results of two function calls, and the act
of adding two variables does not affect how the contents of those variables
change. Similarly, an assignment expression will modify the left hand side of
the expression in a well understood way regardless of what appears on the
right hand side.

In a concurrent model though, this orthogonality is more difficult to ensure. A simple example is when variables are shared across multiple threads
of control. A race condition can cause the result of an assignment statement
to change due to nondeterministic execution. This results in the constructs
of assignment becoming tied to the constructs used for concurrency, violating
traditional assumptions of orthogonality. This violation arises because opera-

tions such as referencing memory suddenly become dependent on whether or not some other thread wrote to the memory. The existence of the memory reference within a thread body means that memory references and threads are no longer orthogonal — one can impact the other and potentially change its meaning depending on its context. In a single threaded program, this code will act the same all of the time:

```
x = a[1] + a[2];
```

If this is contained within the body of a thread, all bets are off — if another thread modifies the array a[], then this code will no longer have a guarantee of determinism.

```
#pragma omp parallel shared(a)
x = a[1] + a[2];
```

So, while the array reference remains orthogonal to constructs we are familiar with (such as expressions and assignment), it is not orthogonal to the parallel thread body (in this case, using OpenMP. See Appendix A for details on OpenMP). The potential behavior of the array reference now is dependent on the context in which it appears within the program.

This consideration becomes very important, as we will see later in languages such as Co-Array Fortran, when constructs are overloaded to both operate on private versus shared data. For example, the assignment operator = is used in Co-Array Fortran for both traditional local memory assignment and to specify non-local communication operations. It can be argued that this is not a violation of orthogonality, but this is not necessarily true. In the case of Co-Array Fortran (or its PGAS cousins), this violation would in fact be avoided if and only if the runtime system binding parallel processing elements (in a potentially distributed system) can provide exactly the same behavior and semantics as a cache coherent shared memory. Unfortunately, for performance reasons it often does not, which means that different semantics are associated with the = operator in the purely local case versus that where shared data appears on either side of the operator.

5.3.1   Sequential constructs and concurrency

Clever mechanisms are possible that use sequential features in a concurrent environment to make concurrency work in a language not designed with it in mind. Consider the concept of a monitor (previously discussed in section 3.2.4). The monitor combines the notion of classes first introduced in the sequential language Simula with Dijkstra's semaphore. Semaphores proved to be cumbersome to use directly because they required all code within a program that accessed other code or data that required synchronization to properly manipulate the semaphores. For sufficiently complex codes, this was difficult to keep track of and was easily prone to synchronization errors.

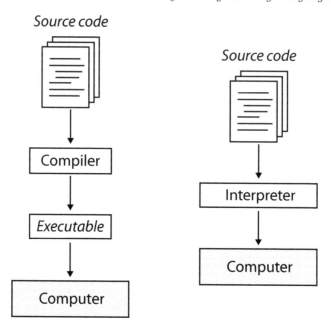

FIGURE 5.2:　Illustration of the stages between source code and execution on a computer for interpreted versus compiled languages.

The insight that led to the monitor concept was that one could encapsulate the sensitive data or code that was to be protected within an object, along with the semaphore used to protect it. By then restricting access behind a set of methods on the monitor object, the locking discipline could be hidden from the calling code and the sensitive data or code hidden. The concept of a class and object-like encapsulation is independent of concurrency, yet it can be used to make concurrency more manageable.

5.4　Interpreted languages

The majority of languages we cover in this text are compiled languages, in which the source code is statically analyzed and compiled into a runnable executable form. Not all languages fall into this category. A popular class of languages are *interpreted languages*. An interpreted language is based on the execution of the program by a special program called an *interpreter*, which sits between the language and the underlying machine providing an abstraction layer above which the programmer works and the program executes. This difference is shown in Figure 5.2.

In theory, a compiled language can be implemented in an interpreted manner, as the interpreter itself is simply an abstraction of the underlying computing hardware. What distinguishes interpreted languages from their compiled cousins is that they are typically designed specifically to take advantage of flexibility afforded to them by the interpreter. This flexibility is often used to provide dynamic capabilities within the language that would be difficult to implement in compiled code. Some of these dynamic capabilities include:

- Dynamic typing: Resolution of the types of objects at runtime based on the runtime behavior of the program.

- Runtime code specification: Introducing new code into the executable at runtime. The `eval` operation in LISP, Scheme, and many scripting languages allows source code to be specified at runtime to be executed.

- Polymorphism: The ability to implement polymorphic code that operates on generic data types without explicitly enumerating the possible types to handle at compilation time.

- Debugging: Interpreted languages are typically easier to debug because the original program code is known to the interpreter, where a compiled program requires additional information to be added to an executable to correlate binary code with the original source.

An additional advantage of interpreted languages is portability. Programs are executed by an interpreter that presents a uniform interface regardless of the actual platform it is running on. The effort for porting to a new platform is largely isolated to the task of porting the interpreter alone. Of course, there are many areas where an interpreted program can be unportable, such as the use of platform specific filenames or an assumption about external utilities available to be invoked.

Some languages that are typically considered to be compiled are actually interpreted languages in disguise. Two familiar examples are Java and the .NET family of languages. In these languages, the compiler produces executables not in machine code, but in a byte code that targets an interpreter known as a *virtual machine*. A virtual machine is simply an interpreter that presents an instruction set (with a binary *bytecode* representation) and execution model similar to a concrete hardware platform, but does not change based on the underlying hardware. The virtual machine essentially provides an interpreter for a platform-neutral assembly language which is targeted by compilers. An advantage of virtual machines is that the low-level nature of their inputs makes mapping to a concrete platform easier, allowing for potentially higher performance. Some virtual machines take this a step further, and provide runtime compilation (*Just-In-Time*, or JIT) of the bytecode to native binary code (or more efficient bytecode representation) which is cached and executed to achieve high performance.

From the perspective of concurrent programming, interpreted languages provide many interesting possibilities. By presenting an abstract machine interface to the programmer, an interpreter is able to adopt a model of concurrency that will be consistent across platforms. The interpreter author is simple responsible for adopting the appropriate concrete implementation of the concurrency model of the language as the interpreter is ported. For example, an interpreted language may provide threads and locks to the programmer with a specific set of interfaces. Different threading libraries provide different primitive functions (or, similar ones under different names or with slightly different semantics). Hiding this from users allows them to achieve portability without concerning themselves with the nuances of each platform their program runs on. Similarly, interpreted languages can provide concurrency constructs even on platforms that themselves do not support concurrent programs. In this case, the interpreter can provide concurrent execution by transparent interleaving to simulate parallel execution in much the same way traditional multitasking operating systems do for concurrent processes.

5.5   Exercises

1. In this chapter, we considered the expression

   ```
   x = a[1] + (a[0] * (b[1] / a[1]))
   ```

   as an example of a simple piece of code that can be adversely impacted by uncoordinated concurrency. Assuming that the array a is shared by multiple threads and likely to be modified frequently, how would you rewrite this single statement as a sequence of statements to behave properly in a concurrent environment?

2. Give an example of a for-loop with a body that operates on elements of an array where each iteration is independent of all others.

3. Give an example of a for-loop with a body that operates on elements of an array where each iteration depends on at least one other iteration. How does this impact the potential for implementing this loop in parallel?

4. Pick two languages, one object-oriented (such as Java or C++) and the other procedural (such as Fortran or C). Compare and contrast them with respect to the cognitive dimensions.

5. Describe the strengths and weaknesses of functional and declarative languages (such as Erlang) for the exploration activity (section 5.2.2) using the cognitive dimensions.

6. Given what we have discussed in previous chapters on concurrency control and explicit threading, discuss each of the cognitive dimensions introduced in this chapter (section 5.2.1) in relation to threading.

7. Give a discussion similar to the previous question, but in the context of message passing instead of threading.

Chapter 6

Historical Context and Evolution of Languages

Objectives:

- Examine the historical evolution of concurrency and parallelism in computing hardware.
- Examine the historical evolution of programming languages with attention towards developments that had an impact on how concurrency was addressed.
- Discuss language features that were motivated or influenced by developments in parallel hardware.

The historic basis for the creation and growth of general purpose programming languages was to provide an abstraction layer above machine code and assembly language to reduce the effort required to construct complex programs correctly that performed well on increasingly complex hardware. As computers grew more capable, the complexity of both the software they could execute and data that the software operated on increased. The necessary advance that increased the capabilities of early computers was the ability to increase their component count. Computers grew rapidly in the 1950s in terms of functional units for performing computation and memory units for storing data. The extremely rapid growth in hardware complexity was felt quite acutely by programmers of the day, and was even codified (albeit informally at the time) by Gordon Moore in what is commonly referred to as Moore's law in 1965 [74].

With this complexity in hardware, manual programming of the machine became too costly, error prone, and time consuming, so high-level programming languages were created to address this need. Practitioners were aware of this trend many years before Moore's paper, as high-level languages gained attention and popularity throughout the 1950s. In addition to taming hardware complexity, users of early computers began to appreciate the problems that arise when existing machines are replaced by newer ones. As new machines were acquired, programs required significant programming effort to port them to the new platforms. It was clear that it was not economically feasible to rewrite software each time a new machine was created.

High-level programming languages provided an abstraction above the ever changing hardware layer, leading to easier porting of program codes and efficient executable creation through compilers. The details about the machine were hidden to a large degree by the compiler, and the automated, formally rooted methods used by the compiler to translate high-level languages into efficient machine code made it feasible to utilize new machines as they came out.

One can speculate that the reason high-level languages became accepted was that computing itself (and, as such, the act of programming) became increasingly widespread. The number of computers increased, and along with it the volume of software that was created, sold, and maintained. High-level languages simply made good sense from an engineering and economic perspective. Looking back through the history of computing, one can observe that unlike general purpose programming, concurrent programming has failed to acquire this mass demand. The vast majority of programmers, until recently, have been focused on sequential programs or programming concurrent systems in a very controlled, restricted environment (such as through the use of database transactions, GUI event models, or tightly controlled operating system code bases).

Operating system and database engine designers are few and far between relative to general programmers, and the field of scientific computing where most parallel computing has occurred over the last few decades has been a niche market. The lack of a widespread demand, and a set of very different and specialized users can be considered a key reason for the lack of high-level *concurrent* languages of the same maturity and caliber as their sequential counterparts. Without a critical mass to drive them forward and sustain the language infrastructure (standards, compilers and tools), they have consistently fallen by the wayside while lower-level abstractions based on explicit threading and message passing have persisted. With the advent of parallelism on the desktop, laptop, and gaming systems, this is quite likely to change.

In this chapter, we will discuss notable languages that made a lasting impact on the programming languages community, specifically in the context of concurrent programming. We will also discuss the evolution of hardware platforms and the role they played in guiding language developments over time. Finally, we will briefly discuss why fully automatic parallelization of arbitrary sequential programs has not yet been achieved.

6.1 Evolution of machines

The original computing machines, dating from ENIAC and MANIAC in the 1940s, were devices built to execute sequences of instructions very quickly (where "quickly" in the early days was hundreds or thousands of instructions per second). Technological limits in hardware caused most of the engineering and design effort for many years to focus on increasing the speed of the processing units and, more importantly, increasing memory and storage capacity. Parallelism was rarely considered, although systems such as the IBM 709 in the 1950s did support multiple channels to memory allowing limited execution of simultaneous operations [1]. Parallelism in the early years of computing was focused on maximizing utilization of the computer given the extremely high costs to procure the relatively specialized and rare machines.

Batch processing systems could keep systems busy to a large degree (assuming there were enough jobs to always have one in the queue) by allowing users to submit jobs to a scheduler that was responsible for executing them for a set of users. In the early days, the scheduler was actually a person who managed programs submitted on punch cards or tapes that were physically delivered to the computing center by users. This person was responsible for setting up the input program and data, executing it, and giving the results back to the user, repeating this process for each job in their queue. Even after the human operator was replaced with automated operating systems that managed batch jobs, fundamental issues were present when computing devices executed one and only one job at a time. The problem with this approach was that the speed of the I/O system was, as it remains, significantly slower than that of the CPU. Jobs would spend large portions of their execution time with an idle CPU while I/O occurred. This meant that as CPUs got faster and I/O lagged further behind, the utilization of the machine decreased.

6.1.1 Multiprogramming and interrupt driven I/O

Two solutions to this utilization problem were invented, both of which brought the issue of concurrency to the attention of operating system designers. These were *multiprogramming* and *interrupt driven I/O*. In these early days, concurrency was solely the concern of the system designer — application writers working at a high level of abstraction programmed as though they were working on a single sequential system, much like many programmers today.

Multiprogramming was a technique for multiplexing a single hardware resource such as a CPU by interleaving the execution of multiple programs by giving each program a small time-slice in which it could execute before being preempted by the system to service other programs. This was sometimes referred to as *time-sharing*. The basic concept here was that programs performing high overhead I/O would be preempted, and the time the CPU would otherwise spend idle could be used to allow other programs to run. The end result was a reduction in the amount of idle time spent by the CPU due to I/O, and a corresponding increase in utilization.

Interrupt driven I/O . was a solution to how high-overhead I/O operations were implemented. In early systems, programs would "busy-wait" on I/O transactions, essentially executing a tight loop asking the I/O subsystem "are you ready yet?" The problem with this is that the time spent waiting on I/O was not truly idle, but spent consuming large numbers of cycles with a sequence of instructions that achieved no useful work. Interrupt driven I/O allowed the system to essentially put programs into a wait state, allowing other processes to execute instead of the busy-wait loop, waking the waiting processes up only when the I/O subsystem causes an interrupt to occur signaling that the I/O transaction had completed and the program could proceed.

This model was sufficient for some time, and provided an abstraction that has an apparent and strong influence on modern operating systems today. Unfortunately, CPUs continued to increase their capability, leading to two new sources of performance problems. First, the main memory of the computer, like traditional storage I/O, became significantly slower to access than the cycle time of the processor. Second, increasingly complex instructions (such as floating point units and specialized operations) required more and more clock cycles to execute a single instruction. This led to many innovations that persisted through various revisions and refinements through today. These include *cache-based memory hierarchies*, *pipelining*, and *vector processing*.

6.1.2 Cache-based memory hierarchies

The introduction of a hierarchical memory based on caches took advantage of a concept we consider quite often in designing concurrent programs known as *locality*. Over time, programmers realized through practice and analysis that given an access by a program of an address in memory, subsequent accesses would have a high probability of occurring on an address relatively close to the first. For example, all programmers have written a loop at some point that walks an array in order performing an operation on each element. Hardware designers realized that they could build very fast memories (relative to main memory) that had a smaller capacity with a lower access time that could exploit locality to hide the higher latency required to access the main store.

The basic concept of a cache is that when an element of main memory must be accessed, the desired value *and a small set of its neighbors* are all retrieved at the same time and stored in the small, fast cache memory. In the high probability event that subsequent accesses will be local to the original, the values would be retrieved or stored to this fast cache memory. By amortizing the high cost of the original access over the full sequence of memory operations, the performance of programs would increase.

Caches were a huge architectural innovation, hence their standard presence on nearly all modern systems. Similarly, pipelining was invented to increase the throughput of the processor to reduce the number of cycles on average required to execute a single instruction.

6.1.3 Pipelining and vector processing

Pipelining took advantage of increased clock speeds and decreased physical component sizes to allow for replication of logic circuits to create a stack of stages much like the stations on a factory assembly line. Vector processing takes a similar approach, but instead of making the assembly line longer, it makes it wider. In the 1960s, the Solomon project at Westinghouse, which led to the ILLIAC IV developed at the University of Illinois, explored the concept of multiple floating point units concurrently operating on elements of an array. This was an early form of *vector processing*. This had a lasting impact on subsequent processor designs, as vector processing and pipelining are present in nearly all modern CPUs.

Interestingly, the development of machines like the ILLIAC IV led to language developments at the same time to make these features accessible to programmers. For example, the IVTRAN derivative of FORTRAN was developed containing "statement forms for expressing parallelism and data layout in array memory." [73] These language methods specifically for dealing with parallelism and data layout issues required for array processing bear a striking similarity to modern issues related to parallelism on multicore and data management for specialized accelerators. Another line of work with similarities to today is the Glypnir language [67], also developed for the ILLIAC IV as a derivative of ALGOL. Unlike IVTRAN, Glypner exposes the underlying details of the machine to the programmer (which, in the conclusion of [67], is considered to be a "major deficiency"), which is quite similar to modern languages like CUDA or libraries for specialized accelerators.

During the same period of time during the 1960s and 1970s, Control Data Corporation developed a line of computers such as the famous CDC 6600 and 7600. The CDC 6600 was based on parallel functional units being present in the central processor, similar to the modern superscalar processors that are used today. The CDC 7600 was a refinement of the parallel functional unit design of the 6600 based on a pipelined approach. Instructions would move from one functional unit into the next, allowing new instructions to start in the units that became freed up in the pipeline. As such, it was able to achieve performance greater than the 6600 by exploiting this instruction level parallelism. The machines designed by Control Data were the direct predecessors to the famous Cray computers, as Seymour Cray left CDC to found Cray Research to create the Cray-1 and its successors, all of which heavily relied upon the vector and pipelined processing demonstrated in the CDC systems.

The key to vector processing is the ability to bring multiple data elements (often contiguous elements of an array) into the processor to perform a single operation concurrently on each element. This eliminates the need to explicitly iterate over the elements, and allows the hardware to perform in one instruction what was previously achieved iteratively with multiple instructions and conditional control flow structure. Similarly, pipelining relies on the ability to decompose complex operations into a sequence of simpler ones such that multiple operations can execute concurrently at different levels of completion.

6.1.4   Dataflow

The von Neumann model of computation , based on the standard processor work cycle of fetching data, decoding instructions, executing them, and writing results back to memory, went relatively unchanged until the mid-1970s. Most architectures up to that point were of that form with minor variations. It had already been recognized at that point that the gap between memory latency and processor cycle times was growing very rapidly, and no end was in sight. The von Neumann model is very sensitive to this gap, as the explicit fetching and commitment of data to memory is a fundamental component to it. If memory latency is significantly longer than the time to execute an instruction, then the processor runs the risk of wasting time idle while it waits for the memory request to complete. A great deal of design work was performed to overcome this issue, resulting in the superscalar processors we see today. Dataflow on the other hand attacked this problem by taking an entirely new approach to computing. Instead of computation driven data movement, dataflow aimed to use data movement to drive computation.

Say we want to build a program that takes two pairs of numbers, computes the sum of each, and multiplies their result together.[1] In the von Neumann model, we can implement this as:

- Fetch a, Fetch b.
- Add a and b.
- Store result $a + b$.
- Fetch c, Fetch d.
- Add c and d.
- Fetch $a + b$.
- Multiply $a + b$ and $c + d$.

Instead, we can look at the operations and see that the multiplication operator takes input that is output from both additions. As soon as those additions complete, it can proceed. Similarly, the addition operations are simply waiting for data. As soon as data is available on both of their inputs, they can execute. We can visualize this as a graph, as shown in Figure 6.1.

---

[1] $(a + b) * (c + d)$.

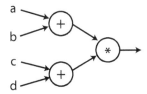

FIGURE 6.1: A basic dataflow graph for the expression $(a + b) * (c + d)$.

In a dataflow system, the operations are executed not based on the program counter being incremented and eventually reaching them, but their input data becoming available. A program is expressed as this set of data dependencies between operations, and the data flow system simply passes data from inputs to outputs based on the data flow graph.

As an architectural change, the dataflow model of computation did not last very long. Dataflow hardware largely disappeared, but the concepts that were learned from programming in a dataflow model persisted. In the scientific computing community, dataflow persisted in the form of dataflow languages. These found some footing on the massively parallel architectures that came soon after the dataflow systems. Unfortunately, hurdles in the efficient implementation of dataflow languages for general purpose programming led the parallel computing community to look elsewhere for a solution. Interestingly, the dataflow programming model has recently found footing in the design of systems for describing and managing business and scientific workflows.

6.1.5 Massively parallel computers

In the 1980s, a new wave of interest in parallel computing occurred based on the idea of making huge numbers (hundreds or even thousands) of processors available within a single machine. These computers were called *massively parallel processor* or MPP machines.

Two branches of the MPP class of supercomputers emerged in the 1980s. One was based on simple processors that carried out many instances of an operation at the same time, each on different data elements. In other words, a single stream of instructions were applied to many individual data elements. Even if each operation was performed relatively slowly, the sheer number of operations executing in parallel during any unit of time would cause the massively parallel system to exceed the performance of more capable systems with fewer processors that couldn't achieve the same degree of parallelism. Notable platforms from this era were the Goodyear Massively Parallel Processor, the Connection Machine produced by Thinking Machines Corporation [53], and systems from the MasPar Computer Corporation.

The second branch of MPP computers to emerge from the 1980s were based on commercial-off-the-shelf microprocessors. Large numbers (dozens to hundreds and eventually in the 1990s, thousands) of microprocessors were integrated tightly with an interconnection network to produce an MPP where each processor ran its own independent stream of instructions on its own local memory. The first notable machine in this class was the Cosmic-cube from Caltech which organized 64 Intel 8086 microprocessors into a single computer. Over time, companies such as Ncube and Intel produced commercial versions of these microprocessor based MPP computers. By the 1990s, this class of machines came to dominate the top of the list of 500 fastest computers in the world (the Top 500 list[2]).

A key feature of both classes of MPP systems is how the processors were connected together. Algorithms to fully utilize these computers had to take into account not only the operations that were performed on each processor, but how data was moved about inside the machine. The interconnection network topology within the machine dictated the paths that data could take between processors. Popular topologies included meshes, tori, cubes and hypercubes, and various types of trees. Application programmers had to carefully optimize their software so the communication patterns implied by their parallel algorithms efficiently mapped onto the network topology.

From a programming perspective, developers needed to express the operations that occurred in parallel across the processors and the movement of data between processors. For the microprocessor based MPP systems, the programmer managed data movement explicitly using communication libraries with little or no support from high level programming languages.

On our first class of MPP machines, those based on a single stream of instructions executing in parallel across the system, data parallel programming emerged as a powerful high-level concept to make the programmers task easier. The CM Fortran language [2] for the Connection Machine was a variant on the Fortran 90 language, which had introduced data parallel features for array computations in the 1990 standard. Communication between processing elements was achievable by using operations to shift elements of arrays in parallel, making the communications operations themselves data parallel. Essentially, all of the processors would execute an operation that would mean "send data to the processor on my right" at the same time. The directions available for movement were dictated by the interconnection topology, and the choice for each processor was dependent on the algorithm.

The concepts in data parallelism that were learned on the massively parallel machines, and the extensions to Fortran that they drove, are discussed further in this text. We will dedicate significant attention to how abstractions based on arrays and data parallelism are a powerful method for concurrent programming (see section 7.1). Their influence can be seen in modern ver-

---

[2]http://www.top500.org/

sions of Fortran, interpreted languages like Python or MATLAB®, and new languages like CUDA or OpenCL.

6.1.6 Clusters and distributed memory systems

In the 1950s and 1960s, the ability to connect computers together was made possible through networking technologies and rapidly became popular. Networking technologies have evolved a great deal over the subsequent decades. A natural insight that was made repeatedly over the decades since networking was invented was that groups of computers could use networks to communicate and coordinate their operations and, in effect, act as one. Unlike MPP computers that use networks internally to tightly couple processing elements to form a parallel computer, clustered systems use general purpose networking hardware such as Ethernet to create loosely coupled parallel systems. Different sources give credit to different products or companies as being the origin of the clustering concept. It is well accepted that the *concept* of coupling computers to work together to perform a single task occurred to many people once networks were available. The best credit for the origin of clusters thus lies with those who created networks and had the insight that computers are quite useful when they are able to communicate.

Clusters gained attention in the 1970s and 1980s when commercial entities began selling products to couple individual machines together using networks to address availability and performance issues. A well known example is the Digital Equipment Corporation (DEC) VAXcluster product intended to couple computers running the VMS operating system [64]. The use of clustered computers became very attractive for parallel computing users with the introduction of the *Parallel Virtual Machine* (PVM) software package [39]. PVM and the Message Passing Interface (MPI) package allow clustered computers to be programmed as though a single virtual parallel computer is present. A single program is written using the PVM or MPI libraries to organize and coordinate the interaction of instances of the program running on the individual computers (typically called *compute nodes*, or simply *nodes*). The software layer hides from the application programmer the underlying networking layers and addressing scheme, providing a convenient abstraction to program with.

Clusters are an important instance of a class of machines known as *distributed memory* machines. In a networked setting, each node can only address its own local memory directly from the CPU. To access data that resides in the memory of a different node, messages must be sent over the network to request and deliver the data from the remote memory into the local memory of the requester. As we will see in section 7.2, message passing itself is a useful programming abstraction for implementing concurrent programs. The reader must take note though that the message passing abstraction itself does not imply that programs using it are always based on the exchange of network-based packets in their underlying implementation. It is entirely possible to compile programs that are based on a message passing abstraction efficiently onto shared memory architectures.

An interesting area of work that is related to clusters (and other distributed systems in a more general sense) is that of distributed shared memory (DSM). DSM systems attempt to hide the distributed memory nature of a cluster under a software layer that gives the illusion of a shared memory amongst the participating nodes. Hardware devices also exist at the networking layer to assist in efficiently using distributed memory machines in a manner closer to that which is present in a true shared memory computer. We refer readers interested in this topic to texts on distributed systems, such as Coulouris, et al. [24], and hardware standards such as the IEEE Scalable Coherent Interface (SCI) or InfiniBand.

6.1.7   Integration

The last significant innovation in architecture was another instance of replication, but this time at the level of the pipeline or entire CPU. The great parallel supercomputers of the 1980s, 90s, and recent times are of this form — including the latest multicore CPUs. Early multicore processors from manufacturers such as Intel, AMD, IBM, and Sun, were essentially equivalent to shared memory computers based on single core processors integrated together onto a single processor die. If one looked at a block diagram of the major components from an SGI, Cray, or IBM shared memory computer from the 1990s, you would see equivalents (often with the same terminology) on a block diagram of a multicore processor.

Similarly, the tiled processors that are being explored by companies like Tilera and Intel bear striking resemblances to the massively parallel architectures of the 1980s. In these designs, many simple processor cores are laid out on a die with an interconnection network such as a mesh provided to connect them together. Programming these processors requires the same attention to the communication topology of a program as was necessary to utilize an MPP system.

Finally, the increasingly popular general purpose graphics processing units (or GPGPUs) are a modern descendant of the powerful vector supercomputers of the past. Instead of being provided as a standalone system, these vector processors are provided today as co-processors for a more standard architecture computer. In many ways, a top-of-the-line laptop today is as though one has a vector supercomputer tightly coupled with a shared memory supercomputer, all together in a tiny package.

6.1.8   Flynn's taxonomy

In 1972, Flynn proposed a classification [35] that distinguished computers on two axes — the number of independent program counters that could be executing at the same times, and the number of data elements seen per instruction by each control flow stream. In each dimension, only two possibilities exist: either singleton ("Single") or sets ("Multiple"). Given two possibilities for each dimension, four combinations exist.

- *Single Instruction, Single Data stream (SISD)*: A single program counter executes on the system, pairing each instruction with at most a single element from memory. Traditional single CPU architectures such as early IBM mainframes and personal computers are of this form. These dominated the computing world, especially in the consumer domain, until only recently.

- *Single Instruction, Multiple Data stream (SIMD)*: A single program counter executes an instruction stream with multiple data elements being paired with each instruction. Early vector computers that could perform instructions on small arrays of data pioneered this concept. Later architectures such as the Intel, AMD, and Power systems exploited SIMD operations for specialized instructions most often geared towards the common linear algebra operations required for multimedia applications.

- *Multiple Instruction, Single Data stream (MISD)*: Multiple program counters executed on a single stream of data. This is a rarely used model, proposed from an academic perspective to address issues of processor redundancy. It is rarely, if ever, encountered in practice.

- *Multiple Instruction, Multiple Data stream (MIMD)*: Multiple program counters executing on multiple streams of data. This model is most commonly encountered in multiprocessor systems of the shared or distributed memory form. This structure is similar to that encountered in clusters and other parallel computers.

The primary contribution of this taxonomy was to provide a common language in which machine architectures could be discussed. Being able to identify an architecture as "SIMD" is informative, as it immediately indicates what performance considerations a programmer must make and where one can expect gains to be achieved. The limitation of a strict reading of Flynn's taxonomy is that the decomposition is based on the program counter level view of the machine. Rarely do we currently look at instruction-by-instruction execution of machines, but instead we think about the execution of instruction sequences or entire programs.

Relaxing this program counter constraint, people began to use the word "program" (or "P") instead of instruction ("I") to categorize parallel programs. For example, it is not uncommon to encounter the term "SPMD," or "Single Program, Multiple Data," when discussing parallelism. SPMD is one of the most common parallel programming models used in practice. SPMD implies that a single executable is run in many concurrently executing processes, and they are each provided their own data sets to work with. Programs based on the common MPI message passing library are usually of this form, in which the same program runs on all parallel processors, but uses its local state (including its rank in the parallel program) to perform its share of work, coordinating with others through messages.

Regardless of whether or not we are discussing programs from the program counter point of view or the overall program, the taxonomy of Flynn and subsequent modifications of it provide a useful terminology for discussing and differentiating between parallel programs in a coarse sense.

6.2   Evolution of programming languages

At the time of writing this text, programming languages have existed for slightly over 50 years. During this period, they have remained a core area of research and development in the field of computer science. Languages have proliferated as the needs of developers expand and change. These needs have been driven both by increasing application complexity and the ever changing hardware landscape that developers are tasked with targeting. The changes in complexity require languages (and their corresponding tools) to assist developers with managing the structure of the code and analyzing various types of correctness properties of the code that are important. In this section we are concerned with looking specifically at the evolution of languages with respect to development of programs that utilize concurrency.

Rarely have languages been created out of nothing, and instead, they have evolved as extensions of existing languages or recombinations of multiple different languages. Thus, it is informative to understanding modern languages by examining how they came to be. This section can be skipped without fear of missing valuable content for the remainder of the book. We encourage readers to refrain from skipping it though, as the historical element of how we arrived in the present from the developments of the past is quite interesting.

6.2.1   In the beginning, there was FORTRAN

Languages for parallel or concurrent programming often grew out of developments in the sequential language community. Therefore we will start our discussion at the beginning with the original widespread programming language, FORTRAN. Programming early computers was very labor intensive. Programs had to be constructed at the instruction level, and manually input into the computer via switch panels or punch cards. Like modern machines, complex operations were built out of many smaller machine instructions. Assembling programs manually, without any automated tools to aid in producing correct code, limited the practical complexity of programs that could be writ-

Listing 6.1: A basic FORTRAN program to compute the sum of an array of numbers.

```
1 DIMENSION X(100)
2 SUM=0.0
3 DO 4 I = 1,100
4 SUM = SUM + X(I)
5 STOP
```

ten. Languages and compilers were the key to making complex programs feasible.

Although "automatic programming" systems existed in the early 1950s to provide abstractions above the machine that were easier to program, they were not compilers or languages in the sense that we think of them today.[3] When it was introduced in 1957, the **FOR**mula **TRAN**slation language for the IBM 704 computer forever changed computing. Programs could be expressed in human readable text that resembled the mathematical notation underlying the computations to be performed. Given that the early days of computing were driven by scientific computation, it is not surprising that the early emphasis was on mathematical notation and structures.

FORTRAN provided a few key abstractions that influenced later languages. It included the notion of a named variable decoupled from its physical address in main memory. FORTRAN also provided control flow constructs for branching and looping, similarly decoupling the structure of the code from the actual addresses that the program counter traversed during execution. By utilizing these, programmers were not responsible for manually keeping track of layout details of the actual program and data store. Instead, they could program based on names and labels that corresponded to meaningful parts of their program. Automated generation of control flow code allowed for the complexity of loops to increase without burdening the programmer with managing the spaghetti of control flow that results from a sufficiently complex program. This automation of the control flow was not perfect, as abuse of the GO TO control flow construct led to excessively complex and tedious control flow structures, resulting in the infamous paper by Dijkstra [28]. Regardless, it *was* progress.

Early versions of FORTRAN also introduced the concept of a subroutine, a block of code that could be called during the execution of a program and passed data via parameters. This was an important advance over simply defining blocks of code that could be executed via a raw jump and return. The abstraction of the subroutine and its parameters allowed the compiler

---

[3]For an excellent discussion of the precursors to FORTRAN, see the articles by Backus [9], Knuth [62], and Knuth and Pardo [63].

to hide the mechanism by which data was made available to the subroutine and how the program counter was manipulated to transition back and forth between the caller and the callee.

Most importantly though was the issue of portability. A program in the high-level language could be moved to a new machine and be expected to run, given a compiler for the new platform. Issues such as those above, related to memory addressing, instruction sets, and control flow, could potentially function differently on different platforms. Different instruction sets could exist with different behaviors, and register counts could vary — but the program would run with significantly less effort in porting than one written at the native machine code level.

FORTRAN was not the first language to adopt concurrency constructs in its official standard. In fact, it took until 1990 before a version of Fortran was standardized that included data parallel features.[4] Subsequent revisions of the language in 1995 and 2008 have adopted further features specifically for concurrency. On the other hand, extensions of Fortran for parallelism have existed for most of the lifetime of the language. As mentioned briefly earlier, the version of FORTRAN for the ILLIAC IV, IVTRAN, [5] was an extension of the existing FORTRAN language to assist programmers with the specialized hardware the machine provided. IVTRAN introduced a modified DO-loop called a `DO FOR ALL` loop, which specified that assignment statements based on the index sets of the loop were to be performed in parallel.

6.2.2   The ALGOL family

Evolution beyond FORTRAN occurred for many reasons, most dominant being the need to manage increasingly complex programs, and the desire to build languages more conducive to efficient automated compilation. A few years after FORTRAN was introduced, a new language called ALGOL (for *ALGO*rithmic *L*anguage) was invented and brought the notion of *structured programming* to the language landscape [79, 89]. Structured programming aimed to tame codes that became needlessly complex to manage as they grew larger. The original ALGOL 60 language was a significant milestone in early language development, with the introduction of familiar constructs such as *if-then-else* and the *for*-loop. It also was the first major language to use a well defined formalism for describing the language syntax through the Backus-Naur Form (BNF). The language was not without flaws, although this is not surprising given the fact that the language was the first to explore many concepts (see Knuth [61] for details on these early issues).

---

[4]Fortran 90 and subsequent revisions of the language also adopted the modern convention of only capitalizing the first letter of the name.

[5]For any who haven't noticed why IVTRAN was given that name, just recall what the Roman numeral IV represents, and say the name out loud.

Listing 6.2: A sequence of ALGOL statements.

```
X=1;
Y=X+1;
W=X-1;
Z=Y*X*W;
```

Surprisingly, refinements of the ALGOL language in the 1960s addressed more than these fundamental shortcomings in the original definition. The designers of ALGOL 68 decided to directly address the problem of expressing parallelism within programs in this revised version of the language. In ALGOL 68, the *collateral clause* concept was added to call out portions of statement sequences that could be reordered or executed concurrently without detrimental effects on the output of the program. Furthermore, the language provided synchronization mechanisms through variables of the **sema** data type[6] corresponding to Dijkstra's semaphores. Concurrency became a topic of interest in the computing community during the 1960s as the limits in utilization and performance of single process batch systems was becoming apparent, especially with a growing gap between the performance of computational elements and data storage such as memory and tape.

ALGOL 68 collateral clauses and semaphores

The collateral clause was quite simple. In ALGOL, statements were ordered in sequence by semi-colon separators, much like the modern languages derived from C and C++ (which they themselves inherit from ALGOL). Consider the sequence of statements in Listing 6.2.

On examination, we can see that the second and third statements do not affect each other — they read from the same input, and write their results into different variables. So, there is no harm in either reordering them arbitrarily or executing them at the same time. We can express this via a collateral clause by simply replacing the semi-colon that separates them with a comma, as shown in Listing 6.3.

While seemingly a trivial modification, the impact is quite profound. Instead of relying on data-flow or control-flow analysis to infer the independence of these operations, the programmer can specify it directly! This not only simplifies the task of the compiler, but allows concurrency to be identified in regions of code of complexity sufficient that flow analysis algorithms are unable to infer statement independence. It should be noted that in languages like C where the comma separator can be used to separate statements, there is a required ordering specified by the language. This is to ensure that side-

---

[6]ALGOL types were referred to as *modes* in the papers describing ALGOL 60 and 68.

Listing 6.3: The use of an ALGOL collateral clause. Note the use of the comma to separate the second and third statements.

```
X=1;
Y=X+1,
W=X-1;
Z=Y*X*W;
```

effects in statements occur in the sequence order as specified in the program text with the result of the final statement being the result of the sequence. C does not provide collateral clauses, even if it provides a similar comma-based statement separator for some instances of statement sequences.

In many applications that use parallelism, the individual processes must communicate with each other. Often there is some level of synchronization necessary to prevent correctness problems in the implementation of this communication. ALGOL provides a variable mode called **sema** that can be used to implement this synchronization through semaphores. Variables of mode **sema** can be operated upon using the **up** and **down** operators that increment and decrement the variable. The compiler must be notified explicitly when **sema** variables are present within a collateral clause by encapsulating the clause within a *parallel clause* via the **par** symbol. The concept of the *co-begin* found in subsequent languages can be traced back to the **par begin** introduced in ALGOL 68.

Concurrent Pascal and Modula

In the 1970s, languages with more explicit intentions to be used primarily for concurrent programming began to emerge. One of the first, Concurrent Pascal [49], was created for the purpose of "structured programming of computer operating systems."

Interestingly, the dominant focus of concurrent languages in these early days was on operating systems programmers. Only in the late 1970s and early 1980s would significant attention be paid to other application areas. Andrews and Schneider [6] in 1983 attribute this to the advances made by the early language pioneers, and the emergence of increasingly affordable hardware.

Concurrent Pascal extended the existing Pascal language by adding processes and monitors. Unlike the raw semaphores provided by ALGOL 68, monitors provide an encapsulation mechanism for associating shared data and the necessary synchronization primitives to protect it with the operations that act upon it. Monitors couple Dijkstra's semaphore primitive with the class structure introduced in Simula [78].

Modula [92, 93] was designed as a successor to Pascal and draws on concepts from Concurrent Pascal in its treatment of how parallelism is expressed. Instead of explicitly calling out operating system design issues as a motivating factor, Wirth instead made it clear that dealing with concurrently operating peripheral devices was a primary design motivation in Modula.

6.2.3   Coroutines

Coroutines are often brought up in the context of parallel programming because, while not a form of concurrency in their original form, they share some properties with parallel code. Coroutines were proposed initially by Conway [23] in 1963 as a method for managing the complexity of designing COBOL compilers. A coroutine is defined in this 1963 paper as:

> An autonomous program which communicates with adjacent modules as if they were input or output subroutines. Thus, coroutines are subroutines all at the same level, each acting as if it were the master program when in fact there is no master program.

Coroutines do not assume concurrency in their design, as only one coroutine is active at any given time, and control is voluntarily given to a peer routine when the active routine is ready. On the other hand, programs structured as a set of interacting coroutines have much in common with message passing programs. The concept of a coroutine provides a design pattern that can *lead* to concurrency quite naturally even though the coroutines on their own are not inherently related to concurrency. The act of decomposing a monolithic algorithm into a set of interacting coroutines is often a step in the right direction towards decomposing it into a set of interacting processes that can execute concurrently.

The reason coroutines alone are not immediately transferrable to a concurrent execution model is that programming with coroutines does not restrict the programmer from utilizing side-effects or other concurrency limiting techniques in their construction. Coroutines simply decompose the program into distinct units that exchange control over time. Beyond that decomposition, programming with coroutines does not restrict the programmer in any way, including those that can make concurrent execution difficult or error-prone. Furthermore, coroutines are a deterministic method of programming, which does not fit well within the nondeterministic environment of concurrent programming. A closely related, but far more powerful mechanism was introduced by Hoare in the form of Communicating Sequential Processes.

6.2.4   Communicating Sequential Processes and process algebras

One of the significant milestones in the development of concurrent programming languages and techniques was the introduction of the Communicating

Sequential Processes (CSP) language in 1978 by Hoare [55]. As described in the original paper, the premise of CSP was that programs are fundamentally based on input and output abstractions used to connect sequential computational processes together. The language starts with a technique introduced earlier by Dijkstra known as *guarded commands* [29]. Dijkstra defines looping (do-loops) and conditional constructs (if statements) whose bodies are composed of commands (or command sequences) that are predicated with boolean conditions known as guards. Those which have a guard that evaluates to true are eligible for execution, although the order is not specified. This allows for the introduction of nondeterminism in the program, a fundamental requirement for concurrent programming.

CSP takes the concept of a guarded command as the mechanism for defining sequential control flow, a parallel command used to specify a set of sequential processes that execute in parallel, and basic commands for input and output to provide communication between them. Like coroutines, sequential processes interact by providing input and output to each other. Unlike coroutines though, this occurs through abstract communication channels instead of subroutine invocations. The result allows for both concurrent execution of the sequential processes, and nondeterministic behavior.

Many languages were influenced by CSP. One of the notable languages was occam, a language invented in the 1980s to program the INMOS transputer architecture [32]. In occam, programs are built out of a small set of primitive constructs:

1. *Processes*: In occam, a process is defined more generally than a strict operating system process or thread. An occam process is composed of an environment (e.g.: variables) and a set of actions that operate on the environment.

2. SEQ: This specifies a sequence of commands that are to happen in a strict order.

3. PAR: This specifies a list of commands that are to be executed without specifying the order. A PAR block completes when all of the statements it contains complete.

4. ALT: The ALT command is the occam equivalent of Dijkstra's guarded commands, in which a set of processes are specified along with guards that determine whether or not they can execute. When the guards are evaluated, exactly one of those that have guards that are true is chosen to execute. The choice is nondeterministic. ALT is discussed further in section 7.3.2.

5. *Communication channels*: Communication channels can be defined to allow processes to communicate. A process can either send data into a channel, or read data from a channel into a variable.

Listing 6.4: An example of an occam PAR block containing two SEQ blocks that execute concurrently and communicate via a channel.

```
CHAN a
VAR x,y
PAR
  SEQ
    x := 1
    a ! x
  SEQ
    a ? y
    y := y + 2
```

For example, a block of statements that execute sequentially would be specified as:

```
SEQ
  x := 0
  y := x+1
```

The dependency between the assignment to y and the initialization of x means that they must be executed sequentially, so the SEQ block is the most appropriate. On the other hand, some sequences do not contain dependencies and can be executed in any order, possibly concurrently, without problems. To execute statements concurrently, we would simply place them in a PAR block:

```
PAR
  x := 1+2
  y := 3+4
```

These can be nested, allowing a PAR block to be composed of a set of SEQ blocks that themselves execute sequentially internally, but concurrently with respect to each other. Processes can be defined that represent more complex operations (such as functions). Communication is achieved by declaring channels. Channel declaration is very similar to that used for variable declarations, simply using the CHAN keyword before the list of channel names. Sending is achieved by using the ! operator on a channel and expression, such as X ! 1, which would send the value one over the channel X. Receiving data from a channel into a variable uses the ? operator. The statement Y ? V specifies that a value from the channel Y is to be read and placed in the variable V. We can combine these in a simple program in which two concurrent processes execute, with one setting a variable, sending it to the other process which then adds a value to the received result. This is demonstrated in Listing 6.4.

Though the INMOS transputer is largely a historical entity, the occam language has enjoyed a consistent level of interest since its creation, as has its inspiration, CSP. In particular, the development of formal methods for reasoning about concurrency in the form of *process algebras* has frequently made connections back to CSP and occam. Process algebras are a form of language, although by appearance they tend to look closer to mathematical notations than traditional program code. Their purpose is to provide a formal method to describe and reason about concurrently executing processes that interact through communication channels. This formal approach to concurrency is similar to the adoption of Church's lambda calculus for reasoning about functional programs. CSP itself was developed into a formal process algebra shortly after its initial introduction in 1978. One of the foundational works in the area of process algebras was Milner's Calculus of Communicating Systems (CCS), which was developed around the same time as CSP. We will not spend any further time on process algebras in this text beyond this brief discussion.

6.2.5   Concurrency in Ada

The Ada language was developed under contract for the US Department of Defense to build a variety of critical systems for military and commercial applications [57, 10]. It was designed based on experiences with a large number of languages between the late 1950s and mid-1970s, and was significantly influenced by the Pascal family of languages. Introduced in the 1970s, it has evolved through three standards — Ada 83, Ada 95, and Ada 2005. From a concurrency point of view Ada is notable because it includes facilities for concurrent programming in the form of *tasks* with either shared memory or message passing inter-task interactions.

The fundamental part of Ada for concurrency is the task. A task represents an independent thread of control that executes concurrently with other tasks. For example, the code in Listing 6.5 contains two tasks that will execute concurrently with the main body of the procedure.

When the **begin** of the procedure is reached, the tasks to put on the shirt and put on the pants are executed concurrently with the procedure body in which the procedure to put shoes on is called. The procedure body cannot complete until the subtasks are completed, even if the call to put shoes on completes before, say, the task of putting on a shirt. We can also make the procedure call to Put_On_Shoes a task, with the main procedure to get dressed having an empty body. In either case, it should be clear that we would need some form of coordination between the tasks to ensure that impossible (or at least, uncomfortable) situations don't arise — it would be difficult to execute the Put_On_Pants call after the shoes have already been put on.

The primary method for tasks communicating with each other in Ada is through what is known as a *rendezvous*. An Ada task provides what are known as *entries*, which are similar to procedure specifications although they are restricted to occurring within task or protected object declarations. A task provides the body for an entry in what is known as an *accept* statement.

Listing 6.5: Ada code describing two tasks that execute concurrently with the main procedure body.

```
procedure Get_Dressed is
  task Put_On_Pants is
  ...
  end;

  task body Put_On_Pants is
  ...
  end;

  task Put_On_Shirt is
  ...
  end;

  task body Put_On_Shirt is
  ...
  end;

begin
  Put_On_Shoes;
end Get_Dressed;
```

For example, borrowing an example from Barnes [10], consider a simple task that represents a container holding a single integer item that waits for an item when it is empty and waits for another task to retrieve it when it is full.

```
task Container is
  entry PUT(X: in INTEGER);
  entry GET(X: out INTEGER);
begin
  loop
    accept PUT(X: in INTEGER) do
      V := X;
    end;
    accept GET(X: out INTEGER) do
      X := V;
    end;
  end loop;
end Container;
```

Other tasks can access these entries on the `Container` by calling the `PUT` and `GET` entries as though they were procedures, such as `Container.PUT(42);` or `Container.GET(Y);`. When the task reaches an `accept` statement, it waits for a caller to execute the corresponding `PUT` or `GET` call. Similarly, the caller will block until the appropriate `accept` is reached in the task. This is the basic concept behind the rendezvous model, which can be considered to be a form of synchronized producer/consumer interaction between the caller and callee. The arguments to the entries are sent as messages between the caller and callee, and vice versa when data is returned.

In the example above, the task is forced to repeat a loop in which an `PUT` entry must be called before a `GET` can occur. In some programs multiple entries are valid to be accepted at a given time. This is provided through the `select` syntax.

```
select
  accept A() do
  ...
  end A;
or
  accept B() do
  ...
  end B;
end select;
```

This allows either `A` or `B` to be executed, and the choice of which actually occurs depends on which arrives first. On the other hand, sometimes the determination of which should occur is based on the state of the task providing the entry. Guards can be placed on the `select` clauses to accomplish this.

```
select
  when QueueSize > 0 =>
    accept Dequeue(X: out INTEGER) do
      ...
    end Dequeue;
or
  accept Enqueue(X: in INTEGER) do
    ...
  end Enqueue;
end select;
```

The second clause is implicitly guarded with the value true. If a `select` contains clauses that are all guarded with expressions that can potentially evaluate to false, any program execution in which they *all* evaluate to false will result in a runtime error.

Tasks can be used to build synchronization primitives such as semaphores and monitors. They also provide facilities for dealing with timing, such as delaying an operation or timing out if a rendezvous does not occur within a specified period of time. Many of these finer details about tasks are outside the scope of this discussion. The key concept that Ada provides in the context of concurrent programming is that of *rendezvous* synchronization implemented through *entries* on *tasks*.

Ada 95 (and later) also provides entities known as protected objects that implement synchronization on data object accesses instead of between separate threads of execution. For example, a protected object might represent a shared counter. Clearly we would like to ensure mutual exclusion when updating the state of the counter so as to avoid missed updates should two threads simultaneously attempt to access it. Using protected objects provides the programmer with language-level support for ensuring this mutual exclusion. Mutual exclusion is a simpler and less general method for synchronization than the rendezvous model that has been provided by Ada since the original language specification. It is likely that the common occurrence of cases where mutual exclusion is necessary to protect data was one motivation behind the inclusion of protected objects in the Ada 95 standard. Previously, this would have been implemented manually through the use of existing synchronization primitives.

6.2.6 Declarative and functional languages

As discussed earlier, functional and declarative languages are based on high-level abstractions that specify what a program is supposed to do, leaving the question of how it actually does so up to the compiler. In a declarative program, the developer gives up fine grained control over how the program is actually implemented on the hardware in exchange for higher-level abstractions intended to make building complex programs easier by making automated

Listing 6.6: A simple ML program.

```
fun lister 0 i = []
  | lister n i = (i)::(lister (abs(n-1)) i);

val x = [2,4,1];
val y = map (fn i => (lister i "X")) x;
val y = [["X","X"],["X","X","X","X"],["X"]]
  : string list list

val z = map (fn i => (lister i 4.5)) x;
val z = [[4.5,4.5],[4.5,4.5,4.5,4.5],[4.5]]
  : real list list
```

tools deal with tedious and difficult details at the lower level. For example, consider the simple ML example in Listing 6.6.

Walking through the code, we first define a function that builds a list of n elements where each element as the value i.[7] Then we create a list of three integers called x. Finally, we use the built in **map** function to apply the given anonymous function to all elements of the list x and call the resulting list y. The anonymous function that is applied to each element of x simply calls the *lister* function to generate a list of capital "X" strings with the specified number. We follow this with a similar statement to create the list z, but in this case we populate the lists with real valued numbers.

Why is this example interesting? It illustrates what a higher-level language buys us in a sequential context. In no place above do we manage memory, nor do we have to worry about polymorphism in the context of the `lister` function applied for different types of i. We don't need to explicitly put checks in for the recursive declaration of `lister` to terminate when it hits the $n = 0$ base case.

At the same time as the development of FORTRAN and ALGOL, McCarthy developed a language that persists in widespread usage today — LISP [70, 71]. The **LIS**t **P**rocessing language was created and developed in the late 1950s and early 1960s as an abstract syntax for a higher-level language that was never popularized. The syntax of LISP as familiar to users of Common LISP or derivatives of the original language such as Scheme is in the form of *symbolic expressions*, or *s-expressions*. An s-expression is a tree-based structure that captures the abstract syntax representation of a program, very similar to the form that results when other languages are parsed into parse trees and refined into an abstract syntax for compilation and analysis.

---

[7]The **abs()** call is to prevent bad behavior when n is negative.

McCarthy originally envisioned a higher level syntax based on what he termed *m-expressions*, which would provide a syntax closer to ALGOL that would be translated into s-expressions. Users of the basic s-expression syntax of LISP were satisfied with what they provided, and the m-expression syntax was largely set aside and not widely accepted. This early choice based on the preferences of users to focus on the internal notation of LISP instead of a syntax similar to other languages can be considered to be one of the factors contributing to the longevity and flexibility of the language.

LISP is at its heart a declarative functional language. Unlike programs written in FORTRAN or ALGOL, LISP programs were formed primarily by defining functions that operated on input and produced output, where input and output were not restricted to data alone. Data and program code shared the same s-expression representation, meaning that functions could be passed around just like data. The details for making this happen were deferred to the compiler and runtime system.

While LISP persists in current times, along with its descendants like Common LISP and Scheme, it also spawned and inspired a variety of other declarative languages. In the concurrent computing world, LISP and Scheme directly influenced a language explored in the 1980s called MultiLISP [46, 47]. MultiLISP was a Scheme derivative with extensions for dealing with concurrent tasks. The main method for creating and synchronizing tasks was the concept of a *future*. A MultiLISP expression using futures to evaluate multiple expressions concurrently might have the form:

```
(foo (future X) (future Y))
```

In this example, the first subexpression would cause the evaluation of X to begin, and immediately return such that the evaluation of Y could also begin while X is still computing. A future expression evaluates immediately to a placeholder token with an indeterminate value indicating that the consumer of the result will eventually be provided a useful value. A future is essentially a contract or promise to the consumer that a value will be provided, and that the consumer must block until the value is available.

Futures are a very elegant mechanism for creating tasks to execute concurrently, and making the necessary synchronization between the child task and the parent that consumes its value implicit. The programmer is not responsible for manually providing any synchronization. Futures have made their way into parallel languages that are currently under development or in use. For example, the X10 Java derivative includes futures in the language type system [33]. Java itself provides a future-based mechanism in the `java.lang.concurrent.Future` object [41]. Functional languages like the Alice ML-derivative provide futures in the type system [83].

MultiLISP also provided a concurrent expression evaluation construct called `pcall`. Consider the LISP expression (F X Y Z). To call the function F, the expressions X, Y, and Z would be evaluated in sequence. In MultiLISP, one can write (pcall F X Y Z), which will cause F, X, Y, and Z to be evaluated concurrently with synchronization implicitly provided such that F would block pending the completion of the evaluation of the arguments. The pcall construct is very similar to other parallel evaluation statements, such as co-begin or PAR.

Unfortunately, these languages have failed to gain significant traction over their history. An resurgence in interest has occurred with the widespread presence of parallelism with multicore architectures, so the future of this class of languages is quite open. Recent work in the Haskell community has shown significant promise with respect to exploiting parallelism in functional programs [59]. It remains to be seen whether or not this new interest in parallelism and functional programming will be enough to overcome traditional hurdles that limited wider acceptance of functional languages outside of the academic and limited commercial communities.

Dataflow languages

In the 1980s, one hardware implementation of parallelism that briefly became popular was the use of many hundreds or thousands of tightly coupled, simple processing elements. Instead of a few expensive, complex processors, the idea was to achieve performance not by raw single processor throughput but by executing many simple operations in a highly parallel fashion. Machines like the MasPar and Connection Machine were examples of this, with many thousands of simple processors tightly coupled to execute together on problems with high degrees of parallelism. One programming paradigm that was explored to exploit these massively parallel platforms was that of *dataflow programming*. In dataflow programming, the programmer states programs as sequences of operations in which a directed graph can be derived representing the flow of data through the program. Vertices represent operations or actions that occur on input data yielding output data, and edges represent the flow of data between these vertices. An operation has a set of edges that lead to it representing input dependencies, and the operation can execute as soon as data elements arrive along these input edges. Upon completion, operations will provide their output to other operations that are connected via their output edges.

Two notable dataflow languages developed during the 1980s and early 1990s are Id [76] and VAL [72]. We can see an example of a dataflow function in Figure 6.2 as presented in a paper describing VAL. The equivalent VAL function source code is shown in Listing 6.7.

The reason the dataflow model is appealing is that dataflow graphs tend to have large degrees of parallelism present. In the figure, we can see that one addition and three squaring operations are able to immediately start given the three inputs. They can start execution concurrently. Similarly, the addition operations that are a part of the standard deviation computation can execute

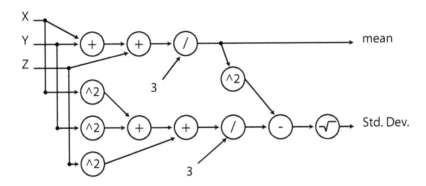

FIGURE 6.2: A simple dataflow graph for a function that computes the mean and standard deviation of three inputs, from [72].

Listing 6.7: The VAL function corresponding to the dataflow graph shown in Figure 6.2.

```
function Stats(X,Y,Z: real returns real,real)
   let
      Mean real := (X+Y+Z)/3;
      SD real := SQRT((X^2 + Y^2 + Z^2)/3 - Mean^2)
   in
      Mean, SD
   endlet
endfun
```

concurrently with the addition and division required to compute the mean. There is parallelism present in the program, and it can be exploited by using the dataflow properties of the program.

To make dataflow parallelism possible, the programming languages in this family were designed to make dataflow graph construction possible. Dataflow programming languages were often based on a core language that was purely functional, implying that side effects were forbidden. Side effects have the unfortunate problem where operations on data at one location in the dataflow graph may be impacted by operations that occur elsewhere in the graph that do not have a direct dataflow dependency. Say two inputs to a function are pointers, and operations that occur on them are independent of each other within the function. From a dataflow perspective, the lack of dependencies within the function means that the operations can execute in parallel. Unfortunately, if the pointers that are passed in point at the same data or components of a single data structure, it is entirely possible that the concurrent execution of the function body may lead to nondeterministic effects. For this reason, side effects must be strictly controlled (or, in some cases, forbidden) in dataflow-based languages.

Unfortunately, most programs require side effects for two reasons: first, I/O is necessary to record the results of a program for later use, and to acquire input data to operate on. Second, some efficient algorithms simply require side effects in the form of direct memory manipulations for either time or space efficiency. The Id language provides facilities to deal with this in the form of both I-structures and M-structures. The core Id language is purely functional, forbidding side effects. In some cases it is necessary to provide limited side effects in which data structures are defined to have values that will be written to at a later time than their definition. This is the purpose of I-structures. I-structures implement what is known as a *single-assignment* restriction. Unlike the purely functional model in which values are bound when variables are declared, I-structures allow the variables to be assigned a value at a later time once and only once. As soon as an I-structure is assigned a value, it is read-only from that point forward. All subsequent read operations will acquire the same value.

Id also provides a mechanism known as M-structures that behave in a similar manner as I-structures, but allow for more flexible usage. An I-structure allows for single assignment with unlimited reads of the singly assigned value. Before assignment, the I-structure is undefined, and then becomes defined forever after once it is assigned to. M-structures on the other hand are undefined until they are assigned to, but become undefined again as soon as they are read from by a single reader. They remain defined only until they are read once, and then revert to an undefined state. Obviously this is ideal for an I/O interface, and very useful for some parallel programs in which a set of tasks listen to a work queue for jobs in which each job should be executed by one and only one process. The reader should see a similarity between M-structures in Id and shared queues that provide synchronized accessors available in more common modern concurrent languages.

Single-assignment semantics have been exploited many times in the development of programming languages that target parallel platforms. The restriction of side effects that they provide make automated parallelization significantly easier. A language influenced strongly by Id and VAL called SISAL ("Streams and Iteration in a Single Assignment Language") was proposed for parallel scientific programming in the late 1980s and early 1990s [38]. A derivative of C called Single Assignment-C (SaC) was created based on the C language with single assignment functional semantics [88].

An intriguing place where dataflow-inspired languages have taken hold more recently is within the business computing community. *Workflow programming* is a popular method for expressing the relationships between business processes and the information that flows between them. Programming of business workflows is very similar to dataflow programming, in that business activities are defined and input/output relationships between them are established. As these inputs are fulfilled, activities can be executed, such as billing, shipping, or initiating orders. The subsequent results of these activities (such as receiving products or payments) can then fulfill the inputs of other activities in the workflow. Real-world workflows have far more sophisticated structures than simple input/output from basic activities, but the basis of the languages shares a great deal with the earlier dataflow ancestors.

Logic languages

Another class of declarative languages originated in the early 1970s with the logic language Prolog. A notable descendant of the Prolog language is Erlang (which we use later in the book and provide a very brief overview of in Appendix B). Erlang is a language receiving significant attention recently due to its elegant parallel programming constructs that can take advantage of concurrency features in modern multicore environments.

The family of languages that derived from Prolog was very different from a programming perspective than the imperative and functional declarative languages. Instead of writing statements that modified state or functions that consumed inputs and produced outputs, logic languages define programs as sets of relations and facts based on formal logic expressions. The fundamental pieces from which these logical expressions are built is logical AND, OR, and implication relations as defined by first order logic.

Like dataflow languages, logic languages were recognized to have an inherent parallelism. Given an expression X and Y and Z, evaluation of the expression to test its truth value requires all three subexpressions X, Y, and Z to be evaluated. As logical expressions are inherently side-effect free, they often contain parts that can be evaluated independently, and therefore in parallel. This fact was exploited in languages like Parlog [22], and gained some attention in the 1980s and 1990s.

Unfortunately, two factors deterred their wider use. First, the hardware landscape changed and did not trend in the direction of massively parallel systems of simple processors that were ideal targets for programs written in these languages. After the Thinking Machines platforms and the MasPar systems, the industry moved more in the direction of lower processor count collections of more general purpose processors in shared memory configurations (such as the IBM SP and SGI Power Challenge and Origin systems), and later towards the clusters of off-the-shelf workstation processors that are commonplace today. Second, these languages were largely incompatible with existing sequential languages. Programmers did not have an easy way to write programs in logical languages to reuse massive existing code bases written in Fortran or C.

The fact that adoption of these new languages would have required a whole-sale rewrite of existing codes made them unattractive from both a commitment and economical point of view. The gains that programmers and users could have made by moving to them were simply too small to warrant the mass recoding of legacy codes in the new languages and programming styles. The cost to move to them would have been more expensive than the effort required to stick with familiar languages with less elegant mechanisms for parallelism. Furthermore, these less elegant mechanisms such as message passing and explicit threading layered on top of existing languages did not prohibit codes that had already been written from being reused.

Moving to declarative languages, either functional, logic, or dataflow-based would have required large recoding or wrapping efforts that were not considered feasible in most cases due to the effort required. Another dimension not to be ignored was the lack of any guarantee of persistence of the novel languages. Compilers existed and were likely to continue to exist for languages like C, C++ and Fortran. The same was not guaranteed for the new languages developed in the logic, dataflow, or functional families.

This is a theme that has persisted throughout the history of parallel and concurrent language development. The measure of success of a language is not its potential or theoretical merit, but the balance of cost of uptake and the performance and programmability benefit that it provides for real codes.

6.2.7 Parallel languages

As the 1970s and 1980s progressed, parallelism became important in some areas of computing not as a solution to the time-sharing problem, but to increase performance by utilizing concurrently operating processing elements. Scientific computing first explored parallelism in the late 1960s but most high-end machines of the era focused on vector processing instead of more general parallel architectures. As systems grew larger and more complex, particularly with the advent of shared and distributed memory multiprocessors, it became harder to program them efficiently. One of the key issues that languages specifically designed for general purpose, large scale parallelism attempted to address was that of *data decomposition*.

In many of the languages discussed thus far, one of the key discrepancies was related to language support for distributing data amongst the processing elements where concurrently executing threads or processes would act on it. Some parallel languages, such as Cilk (used in later chapters and described in Appendix C), focus on parallelism from the task and control flow perspective with less emphasis on data. It was the sole responsibility of the programmer to write the necessary code to, say, distribute portions of large matrices amongst a set of processors. The choice of data decomposition and distribution is important for performance reasons to exploit locality. To remove the responsibility of distributing data from the programmer (who often did so via message passing), languages were invented that included mechanisms to describe data to be operated upon in parallel at a high level allowing the compiler or runtime to distribute the data amongst the parallel processing elements. Two notable parallel languages that we will discuss here are High Performance Fortran (HPF) and ZPL.

High Performance Fortran

High Performance Fortran was created as an extension of the Fortran 90 language, which itself added useful array syntax to the core Fortran language. The HPF language was recently described in detail by Kennedy, Koelbel and Zima [60], and we will give a brief overview of the language here. HPF was based on previous languages that had extended the Fortran language for parallel computing, such as the CM Fortran language for the Connection Machine and the Fortran D and Vienna Fortran languages developed in the academic community. The HPF extensions were intended to support *data parallel* programming over arrays, which is a very common programming pattern in scientific computing. The HPF language was based on Fortran 90 with the introduction of directives that an HPF compiler could interpret. If the directives were ignored by the compiler, code containing them would simply be compiled as a sequential Fortran 90 program. The use of directives within a standard language such as Fortran or C is also used by OpenMP, for the same reason of allowing a single source code to be compiled by compilers unaware of the directives to generate sequential programs.

To describe how data was to be distributed, the `DISTRIBUTE` directive was used to state how an array was to be split up amongst the parallel processing elements. Possible distributions included *block*, *cyclic*, and *block-cyclic*. In a block distribution, the array is broken into contiguous chunks and distributed amongst the processors. In a cyclic distribution, the array is split into finer chunks which are handed out in a round-robin fashion to the processors. Block-cyclic combines these two concepts, allowing blocks of a certain size to be distributed in a cyclic manner amongst the processors. The three distributions are illustrated in Figure 6.3.

The choice of distribution is often algorithm dependent and can impact performance due to variations in the amount of computation that is associated

Block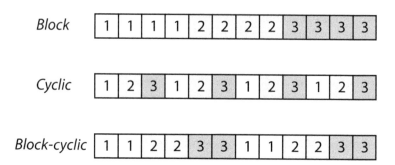

Cyclic

Block-cyclic

FIGURE 6.3:   Illustration of block, cyclic, and block-cyclic distributions of a 1-dimensional array of 12 elements over 3 processors.

with each element of the distributed array. Block distributions typically are appropriate when the amount of computation required for each element of the array is similar. When the computational load per element varies, the cyclic distribution can aid in distributing the load amongst multiple processors instead of potentially concentrating it on a single processor. The block-cyclic distribution allows the programmer to select a balance between the two by tuning the granularity of the blocks that are distributed cyclicly. An example directive for distributing an array X using a block distribution would read as:[8]

```
!HPF$ DISTRIBUTE X(BLOCK,*)
```

Often computations require elements from multiple arrays. Ideally, the elements required in a computation from each of the participating arrays would be colocated at the same processing element. So, in addition to defining the distribution of each array individually, the ALIGN directive was provided to express the relationships between arrays necessary to ensure colocation of subarrays on processors for computations using multiple arrays. An example of a computation requiring this would be the element-wise product of two 2D arrays X and Y. Ideally the ith and jth element of each array would lie on the same processor. This was expressed with the ALIGN directive as:

```
!HPF$ ALIGN X(i,j) WITH Y(i,j)
```

The INDEPENDENT directive was provided for identifying parallelism between loop iterations. If iterations of a loop were known by the programmer to be

---

[8]Readers unfamiliar with Fortran 90 should note that the ! character starts a comment, much like // in C++ and # in some scripting languages. The use of comments to hold directives allows compilers to ignore them if they do not support them.

independent, this directive could be used to inform the compiler of this fact without relying on analysis to infer it from the un-annotated source code at compilation time. For sufficiently complex code in the body of a loop, analysis may not be capable of inferring this.

The final significant language extension was the introduction of the `FORALL` statement, which was later adopted into the mainstream Fortran standard in Fortran 95. The `FORALL` statement provides an alternative to traditional `DO` loops where the content of the loop can be executed in any order with respect to the indices that the loop iterates over. Semantically, the `FORALL` loop is to be assumed to execute all iterations simultaneously, although this may not actually be the case in the underlying implementation. This provides an opportunity to the compiler to emit parallel code to implement the loop.

An example of the `FORALL` statement is the initialization of an M-by-M array to all ones except along the main diagonal where it is set to zero.

```
forall(i=1:m,j=1:m)  X(i,j) = 1
forall(i=1:m)         X(i,i) = 0
```

The iterations of this loop would be performed concurrently, leading to very efficient parallel implementations of the code. A caveat is that the programmer must not assign to the same array element in the body of the loop, as this could introduce nondeterminism in the resulting program since the programmer has no control over the relative order of execution for the loop iterations.

In addition to the HPF directives and the `FORALL` statement, HPF also included a standard library of operations that are frequently used in parallel programs, such as those to discover the number of processors in a HPF program and to scatter and gather data efficiently amongst the processors. While HPF itself did not persist, many concepts that it introduced can be seen in modern languages today.

ZPL

The parallel language ZPL was developed at the University of Washington between the 1990s and 2000s with the goal of providing a portable means of programming parallel computers [21]. ZPL provides an implicitly parallel programming model, in that the compiler takes on much of the responsibility of identifying parallelism and appropriate data distributions within a parallel machine. Programmers who work in ZPL work with a *global view* of data, which means that they work with data structures like arrays in terms of the entire structure instead of the partitioned portions of the structure that are assigned to each processor. The beauty of ZPL being built upon an abstract machine model is that this global view does not dictate shared memory be present — it is entirely possible to use ZPL to program in this global view while allowing the compiler to emit executable code to implement it on a distributed memory machine. When using message passing on a distributed memory system explicitly, we must program with a local perspective of the problem, writing code that explicitly deals with the local view that a processor has of any globally distributed data structures.

It can be argued that the global view programming model makes parallel programming easier [20]. Otherwise, the programmer must manually ensure that the implementations of their parallel program local to each processor implement the global view that they desire when coordinated. The emphasis on a global view of programming is intended to remove this tedium and potential source for bugs.

The fundamental construct in ZPL is the *region*. A region describes an index set over an array. Regions are used to declare arrays and specify the indices of an array that are to participate in an operation. For example, we can define a region R that describes a rectangular index set, declare two array variables over that region, and add their contents (ignoring their initialization):

```
region R = [1..m, 1..n];
var X,Y,Z: [R] float;

[R] Z := X+Y;
```

Regions provide a convenient abstraction to the programmer. They can write code in terms of arrays without concerning themselves with how the actual array data is distributed in a parallel computer. ZPL also provides operators on regions that allow programmers to describe operations that occur on elements of an array in terms of the relative positions of the indices. For example, say we wish to compute the average of the neighbors of each index in an array where the neighbors lie one index to the left, right, above, and below that to which their average value is assigned. These relative positions are defined as *directions*, and can be defined in the 2D case for our region R as follows:

```
direction north = [-1,0]; south = [1,0];
          west  = [0,-1]; east  = [0,1];
```

Now, we can define the operation of computing the average based on these directions over the region R simply as:

```
[R] Y := (X@north + X@east + X@south + X@west) / 4.0;
```

Of course, this doesn't take into account the indices that "fall off" the boundaries of the arrays if both X and Y were defined only on R. We can work around this in ZPL by defining a larger region BigR that contains these boundary elements and declaring X and Y within this region.

```
region R    = [1..m,   1..n];
region BigR = [0..m+1, 0..n+1];
var X,Y: [BigR] float;
```

Given that the code to compute the average is written in terms of indices specified by the region R (as indicated by the syntax [R] to the left of the statement), and the arrays are declared over a region larger than that which the computation is defined means that the boundary values are properly allocated and addressable within the program. A complete ZPL program would also include code that explicitly initializes the boundary indices of BigR necessary for the computation of the average.

When generating a parallel implementation of a ZPL program, the compiler must determine the data distribution that will be used at runtime. ZPL uses the notion of *interacting regions* to determine how indices in arrays relate such that colocation of elements on processors will occur to exploit locality. The motivation for this in ZPL is the same as that which was behind the ALIGN directive in HPF. Interactions also allow the compiler to determine communication and synchronization operations necessary to coordinate parallel processors in implementing the global operation. Interactions are determined based on array operations in statements (such as those using the direction notation shown above), or through the regions used to declare arrays. The distribution of arrays in ZPL seeks to keep them *grid aligned*, defined by [21] as:

If two n-dimensional regions are partitioned across a logical n-dimensional processor grid, both regions' slices with index i in dimension d will be mapped to the same processor grid slice p in dimension d.

In terms of the distributions discussed previously in the context of HPF, we should note that block, cyclic, and block-cyclic distributions result in grid-aligned regions. The ZPL compiler uses a blocked distribution by default.

The key feature of ZPL that readers should appreciate is that programs are written with the global view in mind, such as entire regions, without worrying about the manner by which the array is actually distributed in a parallel computer. Furthermore, the programmer isn't concerned with communications or synchronization necessary to coordinate parallel processors executing the program. Examination of programs written in ZPL (such as the examples available on the ZPL home page[9]) shows that ZPL is very expressive for array-based programming, and the performance results published based on ZPL-programs indicate that a compiler of a global view language can achieve quite impressive results. The ZPL language offers many features beyond those discussed here, and interested readers are encouraged to read more about it in [18] and [26]. The ZPL language is a significant influence on the Chapel language developed by Cray [19].

---

[9]http://www.cs.washington.edu/research/zpl/home/index.html

6.2.8   Modern languages

To round out our discussion of languages, we will discuss a few languages that are currently popular that contain features for concurrency. As of December 2008, the Programming Language Popularity Web site[10] ranks the following languages as highest in popularity based on usage data aggregated from multiple sources:

- C, C++
- Java, C#, Visual Basic
- PHP, JavaScript
- Python, Perl, Ruby
- SQL

Interestingly, data collected based on *discussions* (such as forums, blogs, and news sites) adds some novel languages like Haskell and Erlang to the list. While this data should be taken with a grain of salt, it is interesting to note that it does seem to mimic what one encounters in the world as a working programmer. We see languages traditionally used for applications, such as C, C++, C#, Java and Visual Basic. The Web-oriented languages are represented by Java, PHP and JavaScript. Scripting languages (both for Web and non-Web programming) are represented by Python, Perl, and Ruby. Finally, we see SQL representing the widespread use of databases in a variety of applications.

How do these languages deal with concurrency? C and C++ do not deal with concurrency directly in the language, although many robust threading and message passing libraries are available. Furthermore, many compilers for both languages exist that support the OpenMP parallel programming language extensions. We discuss OpenMP in Appendix A. Java and C# provide similar facilities for thread-based concurrency, using a limited number of keywords (such as `synchronized` in Java and `lock` in C#, and the `volatile` keyword in both) coupled with rich standard library support for threads.

The scripting languages also provide thread modules for writing concurrent code. These threads are typically implemented on top of a lower-level threading library provided by the system, such as POSIX threads. For programmers who are interested in scripting languages like Python and Ruby, it is interesting to observe that additional modules are available from third-parties (that is, developers outside of the core language development community) that make other models of concurrency available. For example, the *candygram* package for Python provides Erlang-style message passing. Similar libraries can be found for Ruby. Perl also provides threading facilities.

Finally, we see that the Structured Query Language (SQL) for database applications also appears in the list of popular languages. SQL is a very

---

[10]http://www.langpop.com/

Listing 6.8: A basic SQL transaction.

```
START TRANSACTION;
UPDATE BankAccount SET balance=balance-5
                   WHERE account=checking;
UPDATE BankAccount SET balance=balance+5
                   WHERE account=savings;
COMMIT;
```

different language from the others in the list, as it is tuned for a specific class of application (database queries). Due to the fact that databases are often accessed by more than one client with potential for concurrent access, SQL includes mechanisms to deal with concurrency. In SQL, concurrency is addressed through the use of transactions. The use of transactions allows for blocks of SQL code to be executed speculatively without modifying the contents of the database directly. Changes made within a transaction are instead written to a log that is associated with the transaction. When the block of SQL within the transaction is complete, the transaction is terminated with a COMMIT statement. This statement attempts to commit, or write, the results of the transaction to the database for permanent storage.

The power of a transaction is that the block of SQL code executed within the transaction does so without acquiring locks, limiting the degree to which other clients would be blocked during its execution. When the transaction is finished and ready to commit, it is possible to see if another client performed some operation that would render the results of the transaction invalid. If this occurs, the transaction can be aborted cleanly without requiring the database itself to be repaired due to the use of the transaction log during its execution. The classic example of transactions used in most literature that discusses them is that of a transfer of money between two bank accounts. We show this in Listing 6.8. Transactions in SQL have been quite influential, not only in the databases world, but in the context of software transactional memory in general purpose languages, which we discussed earlier.

6.3   Limits to automatic parallelization

One of the long sought results in programming languages and compiler research since the early 1970s was the automation of parallelization or vectorization of programs written in common programming languages. If a tool existed that could achieve this, books such as this one would be useless, as there would be no need to understand concurrency unless one was among the few people interested in the design of these mythical tools. The fact that little exists in a general purpose, usable form after nearly 40 years of active research in the computing community should stand as evidence that it is a highly nontrivial problem. Anyone who has written a sequential program and translated it to parallel form may ask the question "why?" though. For many simple programs, the process is fairly straightforward and feels to some degree "algorithmic." The key problem is due to both complexity and generality.

Sufficiently complex programs expose the limits of parallelization algorithms in terms of their time and space complexity. For a complex enough program, compilation may be prohibitively slow or require more space than is available. Similarly, many sequential languages contain features that make parallelization difficult, restricting the applicability of tools for general programs. Unfortunately, some of these features (such as pointers) are extremely common in general programs. In the instances where programs are sufficiently simple and well structured, and specific language features are avoided, compiler writers have demonstrated some level of success at automatic parallelization and vectorization over the years.

Papers on parallelization of DO-loops appeared in the early 1970s [65], preceded by work investigating how one can identify parallelism in programs in the late 1960s [81]. In the early days of supercomputing, where the hardware landscape was dominated by *data parallelism* in the form of SIMD architectures, tools to automatically recognize vectorizable code were investigated and implemented with great success. The fruits of these tools still exist today in modern compilers that target SIMD instruction sets such as MMX, SSE, and AltiVec. Padua and Wolfe [80] provide an interesting and accessible overview of optimization techniques for vectorization of codes for supercomputers in the 1980s.

Many of the techniques for vectorizing codes that were developed over the last few decades are dominated by loop optimizations. As such, their applicability is highly dependent on the amount of loop-based code present in a program. Furthermore, the performance benefit of these optimizations is dependent on the portion of the overall program runtime that the optimized regions of code consume. For numerical codes from scientific fields where loops frequently form the core of the programs, automatic parallelization and vectorization is possible to some degree. Unfortunately, non-numerical codes tend to have a more complex branching structure and many data and functional dependencies between program modules — both of which limit the ability for parallelism to be automatically identified and extracted.

Furthermore, the problem of pointer aliasing (described previously in section 4.4.1), where it is not clear whether or not multiple pointers are pointing to the same location in memory at runtime, severely limits the degree to which analysis tools can identify parallelism. Identifying dependencies is key to analysis algorithms for parallelization, and the pointer aliasing problem makes it unclear whether or not dependencies exist in some circumstances. Compilers most often only have a static view of the program code, and cannot see whether or not aliases do appear at runtime. As such, compilers are forced to make a conservative decision based on their static code analysis and assume that a dependency is possible at runtime.

As we will see in the Chapters 9 through 13, the act of implementing a concurrent solution to a problem requires a deep analysis of the structure of the problem and the relationship between both the steps taken to arrive at the solution and the data elements that are created and modified in the process. Often this structure is not explicitly present in the syntax of a traditional sequential language, requiring effort on the part of the software designer to determine how best to approach the concurrent program design problem.

6.4   Exercises

1. Describe the vector units on your favorite processor, such as the one present in your workstation or laptop. Examples in recent mainstream CPUs include SSE, MMX, and AltiVec.

2. Graphics processing units (GPUs) have recently become popular for programming beyond graphics (GPGPU programming, or general purpose GPU programming). Given that GPUs are based on vector processing, where would they fall in relation to Flynn's taxonomy?

3. Based on the example illustrated in Figure 6.2, draw the dataflow diagram for the following pseudocode:

```
function Foo(A,B,C,D,E,F,G: real returns real)
   let
     X :=  (A+B)*(C+(D/E)+(F*G))
   in
      X
   endlet
endfun
```

4. Explain how you would implement rendezvous synchronization as provided by Ada between threads in Java or C#. What must you implement manually that is not provided by the Java or C# language itself?

5. The OpenCL specification provides a high-level abstraction for programming multicore CPUs, graphics processing units, and other novel architectures with inherent parallelism. Look up the OpenCL specification and describe how concurrency is represented, both in terms of thread creation and synchronization primitives.

6. Given a 1D array of n elements that you wish to distribute over p processors, write the equation used to determine which processor the nth element is assigned to in a block, cyclic, and block-cyclic distribution. For the block-cyclic distribution, assume the block size is $r < \frac{n}{p}$.

7. Discuss how programmers deal with data decomposition in parallel programs when using a simple message passing abstraction without shared memory. Discuss the issues that arise both from a performance and programmability point of view.

8. Repeat the previous question for shared memory threaded programs instead of those based on message passing.

9. Discuss how coroutines relate to concurrency. What aspects of coroutines differ from concurrently cooperating threads?

Chapter 7

Modern Languages and Concurrency Constructs

Objectives:

- Look in-depth at language features for exploiting parallelism through:
  - Array abstractions.
  - Message passing syntax.
  - Control flow constructs.
- Discuss the functional programming model in the context of concurrency.

The basic goal of this text is to introduce readers to the motivation and definition of language level constructs for concurrency that have been designed, studied, and found their way into languages that we can use today. What does this mean? In familiar languages, users employ high-level constructs constantly. Loops are a common example, and are used by defining blocks of code with corresponding control flow constraints. Programmers are not responsible for coding up the sequence of tests and jumps that actually implement the loop. They use the high-level construct for a `while`-loop or `for`-loop, leaving the details of implementing and managing the underlying control flow to the compiler. Similar high-level constructs exist to support operations such as `switch` statements and recursive procedure calls. Languages also provide abstractions of types and structures to allow programmers to work with complex data at a higher level of abstraction than the raw bytes in memory or on disk. The compiler assists with managing the actual layout in the computer and provides various checks to ensure that operations on the data conform to behaviors defined by the language standard (such as properly converting representations when mixing floating point and integer arithmetic).

It is a logical extension then that similar high-level constructs should be available for expressing types and structures that are important within concurrent programs. If one wants to execute a loop whose body is executed by multiple processes in parallel, a construct at the abstraction level of the for-loop should exist that states what should be executed in an iterative fashion while leaving the distribution of the work over multiple processors to a compiler or runtime system to decide. The abstraction is provided to separate the specification of what is to be done from how it is to be performed, providing sufficient information to the compiler in order to ensure that the resulting executable code conforms to the requirements of the programmer, such as the order and dependencies of operations. Similarly, data types should be available to represent data abstractions that the programmer works with at a high level, decoupled from the details pertaining to how the actual data is distributed amongst the processing elements of the parallel system.

In this chapter, we introduce constructs that have been invented for this very purpose. We hope that by understanding these constructs, programmers will be better prepared to use the tools available to them in current and upcoming languages where concurrency is a key aspect of the program design and implementation process. Unlike the previous chapter where we looked at a few instances of concurrency being addressed at the language level over the history of language development, in this chapter we focus on constructs that are available in languages today.

7.1   Array abstractions

One of the easiest ways to introduce parallelism into a language is to take advantage of the inherent parallelism often present in programs that employ arrays. Typically, computations are performed on arrays using loops in an element-by-element fashion. It is often the case that the result of an operation on one element of the array does not depend on the result of the same operation on a different array element. The same result would be computed regardless of whether these element-wise computations are performed serially or in parallel. For example, taking the absolute value of each element in an array, or multiplying them by a scalar constant, does not depend on the order in which the operation is performed. Similarly, many matrix operations can be built out of independent sequences of array operations on the vectors that comprise the rows and columns of the matrices.

The place where this breaks down is when the operations on the array elements depend on elements at different locations in the array. For example, a simple transformation is to shift all array elements one location to the left and assigning the last element to zero, as shown:

```
for i=1 to (n-1)
  a(i) = a(i+1)
end
a(n) = 0
```

Clearly there is a loop order dependence in this example; for if `a(2)` were assigned a value before `a(1)`, then element `a(1)` would end up with the wrong value, as it depends on the *old* value of `a(2)`. This is related to the discussion presented earlier when we introduced the concept of dependencies and their relation to concurrency. In loops, we must take into account dependencies between iterations. Fortunately, parallel array operations exist that are able to deal with these dependencies when they have regular patterns between indices or iterations. An example of such a pattern is a computation on each element of an array being dependent on the same neighbor relative to each index (such as `a(i+1)` in the example above). In this section we discuss array-based abstractions that can be used to exploit parallelism in array computations.

The development of array-related constructs resulted directly from the needs of scientific computation, specifically to support solutions to problems based on linear algebra operations and the solution of partial differential equations (quite often via linear algebra). Similarly, graphics coprocessors share an architectural lineage that traces back to early vector computers since graphics algorithms are typically built on fundamental linear algebra computations. Early vector computers like the Cray 1 were specifically designed to speed up array-based computations. Although arrays play a prominent role for programs in scientific computing and graphics, the use of arrays is ubiquitous in programming. It is easy to imagine that similar abstractions would be of use in processing other sorts of arrays, such as strings or files. Thus, it is worthwhile for all programmers to learn and appreciate the concurrent properties of array-based computations.

The key to making array abstractions possible in a language is the availability of arrays as first-class citizens. Common languages such as C do not provide richly defined arrays in the language. In C, the best the language provides are pointers to memory and syntactic sugar that converts index notation into a memory offset used when dereferencing the corresponding pointer.

Syntactic sugar is a term that the programming languages community frequently uses to describe syntactic constructs that allow the programmer to concisely write code that is equivalent to a potentially more verbose syntax. For example, given an N by N array called `X`, it is easier for a programmer to say `X[i][j]` than to say `X[(i*N)+j]`, even though the former may be translated into the latter by the compiler. Syntactic sugar allows for languages to be designed with a small core surrounded by syntactic forms that are easy to write and have a well defined translation into the core language.

Syntactic salt is another useful term, though less popular. Syntactic salt describes syntactic features that help prevent programmers from writing incorrect code. Explicit scoping blocks present in most languages via curly braces or END keywords is a form of syntactic salt, as they force the programmer to explicitly define block boundaries. Languages like Python that use whitespace for scoping lack syntactic salt present in other languages, leading to potential errors when multiple people with different editors and whitespace settings interact. In concurrent computing, explicit locking or definition of atomic regions could be loosely considered syntactic salt.

An array in C does not contain any information about its length, forcing the programmer to maintain a secondary variable that is used to keep track of the amount of memory contained in the array. For example, a routine that computes the sum of an array of double precision numbers and returns the corresponding scalar value would have a signature similar to:

```
double sum(double *X, int n);
```

Similarly, if one wishes to build algorithms based on sparse arrays or arrays with unusual shapes or index sets, the logic and metadata required to support them must be provided by hand. As such, the C language limits the degree to which language constructs other than individual element indexing can be represented. In C, such abstractions must be written manually as they don't exist in the language. As a result, a compiler is limited in the degree to which parallelism can be easily identified and exploited based on the definition of the language itself.

7.1.1 Array notation

The use of array notation traces its origin to the early days of programming languages with the APL language. Similarly, hardware support for array (or, vector) notation was explored in the 1960s and 1970s, popularized with the introduction of the Control Data Corporation and Cray supercomputers. Vector parallelism is a form of SIMD parallelism (from Flynn's taxonomy, discussed in section 6.1.8). The term "vector notation" derives from the original introduction of this technique in the context of computations on numerical vectors. For example, the familiar dot product of two vectors $\vec{x}$ and $\vec{y}$ is defined as

$$d = \sum_{i=1}^{n} x_i y_i.$$

The element-wise product is clearly embarrassingly parallel, as each pair of elements is multiplied independent of the others. A summation of these pairwise products is then performed, which can be efficiently achieved in logarithmic time in parallel. What starts as a loop:

```
d = 0.0
for i=1 to n
   d = d + x(i)*y(i)
end
```

can be rewritten in array notation with an intrinsic[1] sum reduction operation more concisely as:

```
d = sum(x*y)
```

This notation has two benefits. First, it is more concise and closer to the mathematical notation $\vec{x} \cdot \vec{y}$. Second, the loop becomes implicit. We should note that the product of arrays in the notation above is not a product in the matrix multiplication sense, but the element-wise sense (also known as the *Hadamard* product). In the original code with explicit looping, the compiler is responsible for inferring that the loop corresponds to a single higher-level array operation (sum of element-wise products), and subsequently converting it to an efficient machine implementation that takes advantage of the inherent parallelism of the loop. Although compilers have become adept at automatically vectorizing loops this simple, it is easy to build code that is far more complex where the compiler is unable to infer the vectorizable structure of the loop. By making the loop implicit via a rich array syntax, compilers require far less analysis to identify the potential for parallel execution.

Note that the argument x*y to the sum function is an array-valued expression resulting from the application of the binary operator times to the two arrays x and y. Many of the common unary and binary operators used with scalars can also be applied to arrays and for binary operators, to mixed scalar and array operands. These operators include, the mathematical operators, +, -, *, /, and the logical operators, <, >, ==, for example. With array notation, the operators are applied to each element of the array and the resulting value is stored in the corresponding element of the result array. For example the statement,

```
C = A + B
```

calculates the pairwise sum of elements of arrays A and B and stores the result in the array C. Of course, for this to make sense these arrays must all have the same shape. The *shape* of an array describes the number of valid indices in each of its dimensions. Two arrays have the same shape if they have the same number of dimensions and the size in each array dimension is the same.

This is obvious from considering the equivalent code expressed as a loop, assuming that the arrays are one dimensional,

---

[1]The term *intrinsic* is borrowed from the terminology of Fortran used to define functions that are part of the language and can be implemented efficiently by the compiler for specific architectures. These are sometimes referred to as the *standard library* of other languages.

```
for i=1 to n
  C(i) = A(i) + B(i)
end
```

If the arrays didn't all have the same shape, the common loop bounds assumed by the limits of the loop iterate i couldn't be applied to all of the arrays. Applying this loop to arrays with incompatible shapes would likely result in an out-of-bounds error.

It is important to note that there is no implied order to the evaluation of the operations on array elements with array notation. Thus there are no loop iteration dependencies and the equivalent implementation using explicit loops, shown above, could just as well have run backwards; the results would be the same. This absence of iteration-order dependence is important because it allows a compiler to parallelize the separate elemental operations within the loop.

Array semantics and elemental functions

In languages where array-valued expressions are supported, they are often semantically defined to return a temporary array. It is *as if* the result of evaluating the expression A * B was placed in a compiler-allocated temporary array tmp defined by the loop:

```
for i=1 to n
  tmp(i) = A(i) * B(i)
end
```

Compilers are easily able to identify simple array expressions and eliminate the need for an array temporary. For example, the statement C = A+B would likely be implemented by the compiler as the sum of A and B being placed directly into C without an intermediate temporary for the sum being required between the computation of the sum and the storage into C. But what about expressions involving functions that return arrays? In this instance the compiler may not be able to avoid the creation of a temporary array. Consider the statement,

```
D = A - return_array(B,C)
```

If the compiler doesn't have access to a form of the function return_array that it can manipulate or analyze (such as when it exists in a pre-compiled library), it will likely have no other choice other than to use a temporary variable to hold a copy of the array returned by the function in order to perform the subtraction operation. Programmers must be aware of this implementation detail because it can have a significant hidden impact on the resource requirements of their programs. If one is building a program where memory constraints are a design requirement, then care must be taken when using array notation to avoid expanding the memory footprint of the program beyond that which is explicitly defined by the programmer. This expansion can occur due to hidden temporaries created by the compiler.

In order to get around this problem, modern languages employing array syntax can introduce the concept of an *elemental function*. An elemental function is function defined in terms of scalar quantities, but that can be called with array arguments as well as scalars. A simple example of an elemental function is the binary - (subtraction) operator used in the example above. Subtraction is defined in terms of scalar operands, but for languages employing array notation, the - operator can be used with arrays operands as well. In this instance, the operator is applied to each element of the array, one element at a time, and the result stored in the appropriate location in the result array.

Likewise, users can define elemental functions in terms of scalar parameters and then call them with array arguments. If the function `return_array` had been implemented as an elemental function, then the compiler could effectively replace the example above with the loop,

```
for i=1 to n
   D(i) = A(i) - return_array(B(i),C(i))
end
```

thus removing the need for an array temporary.

Array notation is conceptually easy to think about and to use in programs. It results in code that is easy to read and to understand and is usually close to the original mathematical description of the algorithm being implemented. In addition it is a great feature for compiler writers looking to identify parallelism within program code.

7.1.2   Shifts

Array notation itself is insufficient for many algorithms, as often a computation does not take place on the same index across all elements of an array. Consider a simple algorithm where an array contains a sequence of elements (a sampled audio signal for example), and we want to look at how these values change in time by examining the difference of each element with its successor. This could be written as:

```
for i=1 to (n-1)
   diff(i) = x(i) - x(i+1)
end
```

From a whole-array perspective, this is actually a simple operation where the last $n-1$ elements of array x are subtracted in element-wise fashion from the first $n-1$ elements. In essence, we are subtracting the array with a new version of itself that has been shifted by one element, such that each element is aligned with its successor. An operator `shift` that performs this operation can be used, where the parameters are the array to be shifted and the amount

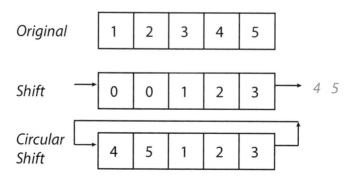

FIGURE 7.1: Illustrating a non-circular and circular right shift by two elements of a simple 1D array.

each index is to be shifted by. Thus, if we wish to align the $i + 1$th element with the ith, we would shift the array indices by -1 such that index $i + 1$ would appear in the ith position of the shifted array.

```
diff = x - shift(x,-1)
```

One subtle difference between this code and the original explicit loop is that in the loop, `diff` had $n - 1$ elements, while the shift-based version has n. Why is this? First, the entire array `x` appears as the left operand for the subtraction operator, so the right hand side of the assignment expression will contain n elements. Second, the shift operator returns an array with the same size as the original. The details regarding what happens at the boundary of the shift are often a property or parameter of the shift operator itself. For example, sometimes a language provides both a circular and non-circular shift operator. A non-circular shift will discard values that "fall off" the end of the array in the direction that the shift took place, filling in the empty spaces created on the other end with some constant (such as zero). A circular shift will simply move the elements that fell off the end back to the other end. The difference between these types of shifts is shown in Figure 7.1.

The power of using shifts is that programmers can express operations in terms of the entire array *and* they are able to encode the equivalent of regular index arithmetic. If we have a shift operation that operates on higher dimensional arrays (such as the Fortran 90 `CSHIFT` circular shift intrinsic, which takes the dimension to perform the shift along as a third argument), we can write a simple statement that replaces each element of a 2D array with the average of its neighbors:

```
x = (cshift(x,-1,1) + cshift(x,+1,1) +
     cshift(x,-1,2) + cshift(x,+1,2))/4.0
```

The first argument to the `cshift` operation indicates the array to shift, the second argument indicates which direction to shift the indices, and the third argument indicates which dimension we wish the shift to be applied to. One could also express this computation with *seemingly* equivalent code using loops:

```
do j=1,n
  do i=1,m
    x(i,j) = (x(i+1,j) + x(i-1,j) + x(i,j+1) + x(i,j-1))/4.0
  end do
end do
```

However, notice that the expressions employing loop indices `i-1` and `j-1` are referencing array elements that have *already* been updated in previous iterations of the nested loops. Thus the two code segments are *not* equivalent. In the version using shifts, we take advantage of the assumed temporary that the intermediate results of the subexpressions on the right hand side of the assignment are stored into to avoid destructive updates to `x` itself until the entire expression on the right hand side has been evaluated. The advantage to the programmer in this case is that the use of shifts versus the explicit indexing in nested loops eliminates the need for the programmer to perform double-buffering or other manual storage of intermediate values during the execution of the loop. For example, if we wish to replicate the behavior of the shift-based code in our loops, we would require a temporary array `tmp` that has the same shape as `x` to store the values during the loop. We would then copy the contents of this temporary array into `x` after the loops complete.

```
do j=1,n
  do i=1,m
    tmp(i,j) = (x(i+1,j) + x(i-1,j) + x(i,j+1) + x(i,j-1))/4.0
  end do
end do
x = tmp
```

One advantage of the implicit temporary that occurs when using array notation is that the programmer is not responsible for allocating and deallocating them in cases such as this. The compiler will emit code necessary to perform this allocation and deallocation transparently to the programmer.

7.1.3   Index sets and regions

Another technique for writing code using array notations employs *index sets*. In most languages, programmers are restricted to writing statements like `x[i] = 4`, in which one and only one element of the array is accessed at any time. A language that supports index set array notation allows a set of indices to be used to address many elements of an array at once. In our previous discussion of the ZPL language, we saw how the region abstraction allows for programmers to use index sets for this purpose. We will see here that languages that support array syntax often also support the use of index sets in array expressions.

We commonly construct index sets by using the colon "slice" or "section" notation in mainstream languages like MATLAB® and Fortran. In MAT-LAB[2] we can address a subset of elements in an array by writing:

```
x(begin:stride:end)
```

Say we want to set all odd elements of an n-element array to zero, and all even elements to 1. We can write this concisely in this notation based on index sets using MATLAB conventions as:

```
x(1:2:n) = 0
x(2:2:n) = 1
```

With this notation on higher dimensional arrays, we can express index sets that represent stripes of the array or checkerboard-like patterns. Recalling our earlier example of computing the difference between adjacent elements in a 1D array, we can write this with index sets as:

```
diff = x(1:n-1)-x(2:n)
```

We should observe that this is different than the version of this operation based on shifts. In the example using shifts, the operands of the subtraction operator were both of length n. In this case though, both operands are of length $n - 1$. Using index sets to express which elements participate in the expression eliminates (at least in cases like this) the issues of boundary values that we needed to worry about when employing shifts.

In ZPL, which supports the notion of index sets via regions, we do not need to define the index set explicitly within an expression. Instead, we can define the regions once and indicate which region an expression takes place over by referring to the region definition. The reader is referred to our previous discussion of ZPL to see examples of this.

7.2   Message passing

One of the most widely used models for parallel programming, especially in the scientific computing community, is the message passing model. The standard library that is used in practice is the *Message Passing Interface* (MPI),

---

[2]In Fortran, the end and stride parameters are reversed, e.g., x(begin:end:stride).

which we discussed and provided examples of earlier in section 4.2.1. The reasons for the widespread use of message passing are many. It is a mature model that has been actively used since the 1970s on many different parallel platforms. Message passing has often aided in portability, as it does not assume anything about the underlying hardware. Programs written using message passing abstractions program to the abstraction and achieve portability by using implementations of the abstraction on different platforms. A message passing program written in MPI can target MPI implementations that utilize shared memory for large SMPs, or network communications for distributed memory clusters. All that is required is that an MPI implementation be made available that maps the abstractions of MPI to the appropriate platforms. Portability of this sort is not found with other parallel programming models in a production sense, although researchers have investigated making models such as OpenMP available for distributed memory architectures in addition to the more commonly targeted shared memory platforms.

Beyond MPI (or its peers), there are languages that also provide message passing primitives at the language level. It is often very straightforward to build a program where one constructs a set of concurrent threads of execution that interact using well defined protocols of data transfer. Thus, message passing is an attractive approach to concurrent programming. The problem with message passing is that the explicit movement of data defined by the programmer limits the ability of compilers to analyze and tune programs to optimize these interactions. There is a tradeoff that programmers must consider when choosing a model: explicit message passing gives programmers detailed control over the precise interactions between threads of control, while higher-level primitives restrict this precision control in favor of abstractions that facilitate compilers in emitting very efficient code.

At the language level, message passing can take on different forms. Below we list three distinct, popular mechanisms for expressing message passing.

- Two-sided messaging in which explicit send/receive pairs are expressed in the endpoints of the transmission. These can be synchronous (similar to the Ada rendezvous) or asynchronous (as in Erlang).

- Direct addressing of data in a remote memory space. This is typically what is known as a *one-sided* operation, in which a reader or writer accesses memory without the direct participation of the process in which the memory resides.

- Remote procedure call or remote method invocation techniques, where messages are passed as arguments to a callable unit of code.

We showed an example of MPI previously in Chapter 4 in Listing 4.1. That example illustrates two-sided messaging using the MPI library calls. At this point in the text, we are concerned with language-level abstractions for message passing instead of those based on libraries.

7.2.1 The Actor model

As introduced earlier, the Actor model is an elegant abstraction above asynchronous message passing. In its basic form it can be implemented on top of standard message passing libraries. It simply requires encapsulation and management of message buffers within threaded objects to provide programmers with the proper abstraction layer to conform to the model definition. A language that is increasing in popularity called *Erlang* is based on the Actor model. We refer readers to Appendix B for a brief tutorial on Erlang, and the more comprehensive book [7] by Armstrong (the inventor of Erlang) that was recently published.

The Erlang language includes primitives for message passing directly within the language. For example, if one Erlang process wishes to send a message to another process (`Dest`) containing the number 42, we can simply write:

```
Dest ! 42.
```

The exclamation mark (`!`) indicates that the result of the evaluation of the right hand expression is to be sent to the process with the identifier `Dest`. This operation does not require that the receiver synchronize with the sender, so it returns regardless of whether or not the recipient has executed the corresponding `receive` operation. On the receiving side, the code to receive the message could be written as:

```
receive
   N -> Foo(N)
end.
```

In this code, the receiver will extract the first message from its mailbox that matches the pattern N and invoke the local function `Foo` with the received data as the parameter. If no message had arrived yet, the `receive` would block.

7.2.2 Channels

The occam language discussed in more detail later in section 7.3.2 provides abstractions for message passing in the form of *channels*. These channels provide an abstract connection between the endpoints of a message passing transmission, and do not dictate a specific underlying implementation by the compiler or runtime system.

```
chan ! src
chan ? dest
```

Languages that provide channels include variable types representing the endpoints of the message passing operation in the form of a channel type, and syntax to express both the placement of data elements into the channel and retrieval of data elements from the channel. In the occam syntax, the operators for sending and receiving are ! and ? respectively, with the channel specified on the left hand side of the operator and the variable that provides the data to send or is the location to store data that is received. In the above example we have a channel variable called **chan**. The first line indicates that the value stored in the variable **src** is to be sent over the channel, and the second line receives a value from the channel that is stored in the variable **dest**. In a real occam program, these operations would likely be separated and reside in independent, concurrently executing tasks.

By appearance alone the notation seems similar to that provided by Erlang, but it is quite different. In Erlang, the process identifier of the recipient is explicitly stated on the sender side, while the receiver simply seeks messages that match one of a set of provided patterns. The receiver does not explicitly state the process from which it expects the message to arrive. If this information is necessary, it must be encoded within the message. On the other hand, occam requires both endpoints of the messaging operation to specify the channel over which the message is delivered.

7.2.3 Co-arrays

An interesting message passing syntax was introduced by Numrich and Reid [77] in a small extension to the Fortran 95 language called F−−, later referred to as Co-array Fortran. The concept behind Co-arrays was to use an array-like indexing scheme for accessing data owned by other processes that do not necessarily share memory with the accessor. Co-arrays are an example of a *one-sided* message passing operation. For example, consider the code:

```
if (this_image() == j) then
  a(1)[i] = 3
end if
call sync_all()
```

This code is executed by a set of processes (or *images* in the Co-array terminology), each of which contain a local array called a which is made available for addressing by remote images by indicating that it is a Co-array.

When a set of processes execute this code, the image with index j will put the value 3 into the first element of the array a on image i. This operation is called one-sided because only the sender executes the operation. There is no corresponding receive operation on the ith image. This is why the synchronization is necessary, to ensure that all processes wait for the message passing to occur before proceeding. When they pass the synchronization point, process i will have the value 3 in its local store.

An interesting, although at time problematic, aspect of using Co-arrays is shown in this example in the need for the `this_image()` guard. If this guard was not present, and the code read:

```
a(1)[i] = 3
call sync_all()
```

then we would see all images execute the same assignment operation, resulting in many redundant messages being sent to the same image setting the same element to the same value. This would be very wasteful of resources. Users of Co-arrays must be conscious of the fact that they are implemented in a SPMD model — the same program runs on all images, so guards are necessary when individual images are intended to execute data transfers instead of the entire set of images.

Co-arrays can also be used to retrieve data using *get* operations, where the Co-array syntax appears on the right hand side of an assignment statement. Syntactically, Co-arrays introduce a new, special dimension to arrays for identifying the image that owns the data. The special dimension used to address remote memory is called the *co-dimension*. The designers of Co-array Fortran chose to make the co-dimension syntactically distinct by using a different bracket character to distinguish Co-array accesses from array accesses on the local component of the Co-array.

Co-arrays are an interesting language feature because they combine message passing with array notation. Given that Fortran is a language intended for numerical computation, most often on array-based data, this is a very natural and comfortable abstraction for programmers who design parallel programs around very large arrays that are distributed amongst processing elements. Recent versions of the Fortran standard have incorporated concepts into the core language to provide concurrency features based on the Co-array concept.

Partitioned global address space languages

Co-array Fortran is one of a loose grouping of languages referred to as the *Partitioned Global Address Space* (PGAS) languages. This family of languages is most often represented by Unified Parallel C (UPC) [90], Co-array Fortran, and Titanium [52] (based on Java). One can argue that other languages fall into the model, but these three are typically held up as the canonical examples of PGAS languages.

The distinguishing feature of these languages is that the programmer is provided with a *global address space*. This means that one process may directly address data allocated by another process. The limitation is that memory that can be addressed must be indicated as globally addressable when it is declared or allocated. For example, a variable that is not declared with a co-dimension in Co-array Fortran is not a candidate for remote addressing.

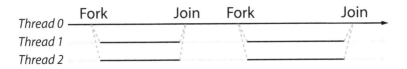

FIGURE 7.2:   An illustration of the transition between sequential and parallel execution in a fork-join control flow model.

7.3   Control flow

Concurrency is often achieved by explicitly controlling the switch between sequential and concurrent execution as control proceeds through a program. For example, a program may start off as a single thread of execution to set up data structures, and then split into multiple concurrent threads that execute on this data. Threads may later rejoin back to a single sequential flow of control, before splitting and joining again in the future. This is illustrated in Figure 7.2. This form of parallelism is often referred to as *fork-join* parallelism for obvious reasons. A rudimentary implementation of this model is available to programmers via the `fork()` system call. In this section, we will discuss different languages and constructs that have used this and related mechanisms to manipulate control flow to achieve parallelism.

7.3.1   ALGOL collateral clauses

One of the earliest instances of this control flow method of exploiting parallelism in a popular language was the *collateral clause* introduced into ALGOL 68 [91]. Consider the following sequence of operations:

```
x=f(1); y=g(); z=f(x); x=y+z;
```

We can identify the dependencies in this sequence by examination to see that the first statement must occur before the third and fourth, and similarly, the second statement must precede the fourth. A diagram showing these dependencies is shown in Figure 7.3.

In ALGOL, one can use collateral clauses to state that the first and second statements can execute in parallel in any order without affecting the outcome

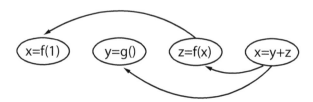

FIGURE 7.3:   Dependencies that impose an ordering on a sequence of operations, exposing some potential for parallelism.

of the program. This is expressed by separating the statements that can be executed in parallel by a comma instead of a semi-colon.

```
x=f(1), y=g(); z=f(x); x=y+z;
```

Similarly, we can find by inspection of Figure 7.3 that we can do this to the second and third statements instead. Either choice will not affect the outcome of the program, but may impact the runtime. For example, if the first statement was fast relative the the second or third, it would be wise to run the second and third in parallel.

This concept introduced in the early ALGOL family has appeared in many forms in languages since then. The basic idea is to allow programmers to provide a list of operations that can be performed in parallel, bounded by syntactic markers for where the parallel execution begins and where the parallel threads synchronize before returning to a single sequential thread.

7.3.2   PAR, SEQ and ALT in occam

Occam is a language that was developed during the 1980s to target the *transputer* platform from INMOS, and was derived from Hoare's CSP and May's EPL [56]. Occam expresses control flow with respect to parallelism via its PAR and SEQ constructs. As we saw earlier in section 6.2.4, the SEQ construct is used to say that a sequence of activities are to happen in sequential order. Say we have the following occam code:

```
SEQ
    x := 42
    y := x + 10
```

This would state that x will be assigned the value 42, and then y will be assigned the value x+10. On the other hand, we may have a set of operations that we wish to execute in parallel, such as one that generates data and one that receives this data from the other to work with. We can use the PAR construct to define this.

```
PAR
  dataGenerator (chan!)
  dataHandler   (chan?)
```

In this snippet, within this block both the dataGenerator and dataHandler processes are started in parallel and handed a channel chan to communicate on. The notation ! and ? on the channel parameter simply indicates the direction that the channel is used by the respective processes, with ! corresponding to outgoing messages and ? to incoming messages. The processes that execute in parallel would then use the channel syntax discussed earlier to read from and write to the provided channel. The PAR construct places an implicit barrier synchronization at the end of the construct. Putting these constructs together, we can achieve interesting control flow involving both sequential and parallel execution.

```
SEQ
  x := 14
  PAR
    foo1(x)
    foo2(x)
  PAR
    bar1(x)
    bar2(x)
  y := 12
```

In this example, we have a sequence that starts with an assignment to the variable x. This is followed by a parallel block in which both foo1 and foo2 are executed. When both complete at the implicit barrier at the end of the PAR block, a second parallel block is started where bar1 and bar2 are executed in parallel. The block again synchronizes on their completion, and continues the sequential execution of the outer SEQ block with the final assignment of a value to y.

While these features are on their own quite useful, a third construct is provided by occam called ALT. This construct is used to provide a set of possible activities to perform, executing only one of them depending on which one meets a set of conditions.

Listing 7.1: A trivial loop with no inter-iteration dependencies.

```
do i=1,100
  X(i) = Y(i) + 10
end do
```

For example, say we have two channels a and b, and want to choose the first that receives a value to assign to a variable x. This can be achieved quite simply.

```
INT x:
ALT
  a ? x
    doSomething(x)
  b ? x
    doSomethingElse(x)
```

Here we see that both of the receives are available, and the ALT construct states that only one is to actually be chosen. The occam language allows for prioritization of which possibilities within an ALT are chosen when more than one was available. Otherwise, the choice is arbitrary.

7.3.3 Parallel loops

One of the most successful areas to exploit parallelism has been at the loop level. Array abstractions that express operations over whole arrays are a specific instance of loop-level parallelism, in which the looping is implicit in the whole-array operation. In general, loops represent repeated instances of a single computation, varying either the location in memory where they operate or some small set of parameters that exist in a fixed location. As mentioned earlier, the work in the early 1970s on automatic identification of parallelism in programs focused on Fortran do-loops. Similarly, sequential compilers have long used optimization techniques such as loop unrolling to take advantage of the regularity of loop-based code to achieve higher performance. How can loops assist us in creating parallel code?

The key to using loops to exploit parallelism is in understanding what dependencies exist between loop iterations and determining the stopping criteria used by the loop in order to terminate. Consider the code snippets in Listings 7.1 and 7.2.

In the first example (7.1), we see that each iteration is independent. Updating the ith element of X requires only the ith element of Y. On the other hand, in the second example (7.2), we see that when $i > 1$, updating the ith element of X requires both the ith element of Y *and* the $(i-1)$th element of X. In the first example parallelization is easy — the loop iterations have no

Listing 7.2: A loop with iteration dependencies.

```
do i=1,100
  if (i == 1) then
    X(i) = Y(i)
  else
    X(i) = X(i-1) + Y(i)
  end if
end do
```

dependencies and can be executed in any order. In the second example, due to the dependency that exists between iterations, parallelization becomes more complex (and in some cases, iteration dependencies can make it impossible).

Some languages have introduced constructs that allow programmers to indicate in which loops they would like to use parallel execution of iterations. Programmers familiar with OpenMP are likely familiar with the parallel loop construct that it provides. Programs that utilize OpenMP primitives can use parallel loops by annotating existing for-loops with a pragma informing the compiler that iterations should be performed in parallel. The advantage of this is that the compiler is not responsible for inferring that the loop can be run in parallel. Instead, the programmer makes this choice. For example, a loop computing the mean of an array may appear as:

```
#pragma omp parallel
{
  #pragma omp for shared(mean) private(i) reduction(+:mean)
  for (i=0;i<N;i++) {
    mean = mean + x[i] / (double)N;
  }
}
```

The first pragma indicates that the following block should be executed in parallel, and the second pragma specifically refers to the for loop. The pragma tags the variable mean as shared amongst threads, i as private to each iteration. It also specifies that a reduction operation must be performed on the shared mean variable to compute the sum of the values computed on each thread to reach the final global value.

A similar looping construct was introduced into the Fortran 95 language in the form of the FORALL statement, first demonstrated in the High Performance Fortran language. The FORALL-loop is similar to a standard do-loop, with the caveat that the iterations are assumed to be valid regardless of their relative order of execution. This lack of inter-iteration dependencies is ideal for parallel implementation. As we have seen, dependencies that are present in programs limit parallel execution. The implied independence of each iteration of a

FORALL-loop makes explicit this absence of dependencies, leaving the compiler great freedom in implementing the loop as a parallel operation.

7.4 Functional languages

It is impossible to write about parallel languages without discussing whole language families that have been long recognized for their potential in parallel environments. Functional languages are a family that has long been of interest in the parallel computing community, examples of which include LISP, Scheme, ML and Haskell.

A specialized functional language that gained large amounts of attention in the late 1980s and 1990s specifically intended to target large-scale parallelism was SISAL. Recently, languages such as Objective Caml, and its successor in development, F# from Microsoft, have gained fairly widespread attention. Unfortunately, these languages have failed to catch on so far in mainstream programming for a variety of reasons. These have ranged from the lack of compilers that can achieve high performance, difficulty interfacing with codes in conventional languages, and the high learning curve associated with moving from the more common imperative style of programming to the functional style. Recent advances in language design and compilation techniques may overcome some of these difficulties. For example, the F# language provides excellent integration with the other .NET languages and the current compiler produces code that performs quite well.

Functional languages are based on defining programs as sequences of functions composed together that act on data. Data is not assumed to be a static entity that is fetched and stored from some memory, but something that is passed around through functions with no explicit control of its allocation and deallocation by the programmer. For this reason, functional programs can be considered (often erroneously) to be inefficient with memory due to the lack of explicit control over memory management as present in traditional imperative languages. Compilers and interpreters for these languages are often quite efficient at analyzing the memory needs of a program during its execution, and the presence of very robust garbage collection algorithms keeps the memory footprint to a reasonable size.

The reason functional languages are of keen interest to people exploring parallel programming is that they are based on an inherently higher level of abstraction. Functional languages are based on the declarative model of programming instead of the imperative. Programs in functional languages express the steps that are necessary to solve a problem instead of the set of lower-level steps that the computer must execute to implement the solution. By avoiding specifying how a problem is to be solved in terms of explicit operations by the computer, the compiler of a functional language has more freedom in choosing which implementation is best suited to the target platform. If this target has parallel computing capabilities, then the compiler has freedom to exploit it in the resulting executable.

In addition to the declarative programming paradigm, functional languages often are very strict when it comes to the use of side effects. In a purely functional program, side effects are prohibited. This means that issues that arise due to side effects leading to correctness problems in concurrent programs go away when side effects by the programmer are not possible. The consequence of this though is that many familiar methods of programming in traditional imperative languages become difficult to express in a purely functional setting. One can argue that this is one of the major reasons that functional languages have failed to gain footing in mainstream, general purpose programming.

7.5　Functional operators

A set of functional operators exist that are often found in the context of functional languages that are powerful for expressing parallel algorithms while not being restricted to functional languages. We can think of these operators as representing algorithmic patterns. By implementing these patterns in languages that we use in practice, we can reap some of the benefits that they provide to users of functional languages. These constructs are quite simple, yet very powerful. They are the *scan*, *reduction*, *map* and *fold* operators. The basic concept behind these operations is that algorithms on large data sets can be expressed as applications of simple functions over these data sets where the application is performed in a very regular manner on singletons or previously processed subsets of the original data set. Though this definition may seem vague, the intent of this section is to make these operations more concrete.

Before we dive into these operators, we must point out one additional feature of functional languages that is important. In a functional setting, functions themselves can be passed around as though they were data. For example, we could have a function f that itself takes a function g as a parameter along with a data element x. The function f could be defined to be the result of applying the function g to some computation on the parameter x. Say x is an integer and we wish to define f as the application of g to x^2. We could define f as:

```
f g x = g (x*x)
```

This function signature (which is valid in Haskell) states that the function f takes two parameters — a function g and a numeric value x. The body of f is composed of the function g applied to x squared. The functional operators that we will look at below exploit this ability to pass functions as parameters.

We start with a data set X that we wish to operate on. Very often the resulting data set Y results from one of a small set of operations on X.

Map

The map operation applies some function f to each element $x \in X$ such that $Y = f(X)$. For example, X may be an array of characters, and Y may be the corresponding array of characters where all characters appear in their uppercase form. So, f would be the function upper() applied to each element. The concept of a map is that the map primitive makes the loop over the elements implicit. Mapping upper() over X is the same as explicitly looping element-by-element over X to produce the elements of Y. The power of the map abstraction is that the loop is hidden from the programmer, thus allowing the language to determine the most efficient method to implement it in the context of other operations happening around the map.

Reduce

A reduction is the application of an associative operator over a data set, where the result often is of lower dimensionality than the input. The simplest example of an operator that can be applied via a reduction is the addition operation $(+)$. A hand coded algorithm that computes the sum of the elements of X would in its most naive form require $O(|X|)$ time complexity. On the other hand, a binary tree of sub-sums could be performed over a set of processors, resulting in $O(log(|X|))$ time complexity for the $O(|X|)$ operations. The reduction operator exists to provide this efficiency for *any* operation that is associative.

Scan

Scan operations are very powerful. They are the basis of *prefix-sums*-based algorithms which can be used to implement sorting, coloring, tree analysis and other algorithms. We start with a binary operator $\oplus$, which can be a variety of operations such as addition, multiplication, max, min, and so on. Given a list of values, $X = (x_1, x_2, \cdots, x_n)$, the prefix sum computation yields a new list Y defined as

$$Y = (0, x_1, \ x_1 \oplus x_2, \ x_1 \oplus x_2 \oplus x_3, \ \ldots, \ x_1 \oplus \cdots \oplus x_{n-1}).$$

The beauty of prefix-sums is that they can be implemented efficiently in parallel with a logarithmic number of parallel steps. Furthermore, it has been shown (most prominently by Blelloch [12, 13]) that they can be used to form the basis of many parallel algorithms. Readers interested in learning about applications of scans and prefix-sums are encouraged to read the papers by

Blelloch as a starting point in investigating this interesting functional primitive.

Fold

The fold operation is very similar to the reduction operator, but with the caveat that the order (left-to-right versus right-to-left) is enforced. Typically fold-left and fold-right operators are implemented if the direction matters to a programmer. A reduction uses associativity of the operator to maximize parallelism by recognizing that the tree of operations has no constraints other than being binary. Fold operations force the tree to have a skewness to either the right or left. This limits the freedom of the implementation relative to raw reductions, as the order of operations again matters. The reason the fold operation remains powerful though is that, like other data parallel operations, the looping of the operation is *implicit*, not explicit. The deferment of how the loop is actually implemented gives the compiler more flexibility in how the code is actually generated. Fold operations also require that the function applied over the list take not only a single element as an argument, but the state accumulated up so far as it has been applied to previous elements.

7.5.1 Discussion of functional operators

The fold, map, and reduce operators are fairly straightforward. For example, consider the string "arachnid" represented as a list of characters. If we have a function that converts a character to the upper-case equivalent, then if we use the map function to apply it to the list, we would obtain a new list of characters representing the string "ARACHNID". Similarly, we may have a function that compares two characters and returns the one that occurs latest in the alphabet. Applying this function as part of a reduction, we would obtain the character 'R' for this string.

Fold is slightly more complex as it depends on the order in which the list is traversed. Consider a simple example in ML in which we create a function that simply prepends an element onto a list. We'll also create a variable x containing the list 1,2,3,4.

```
val x = [1,2,3,4];
fun prepend (elem,lst) = elem::lst;
```

For readers without prior knowledge of ML, the first line simply binds the value representing the list 1,2,3,4 to the variable named x. The second line creates a function called **prepend**, which takes a pair of arguments representing a single element and a list, and returns the element prepended (with the :: operator) to the front of the given list.

Now, we can apply the prepend function over the list x using the left and right fold operations to see what happens. We should note that the pair of inputs to **prepend** are to represent the current element of the list that the function is being applied to (`elem`), and the result of its application to the elements up to this point (`lst`). This accumulated result is often referred to as the *state*. The fold operation takes an initial state (in this case, the empty list), and the list that it is to fold the function over.

```
val y = foldl prepend [] x;
val z = foldr prepend [] x;
```

What results? In the first case, the fold is applied left to right. So, each element is prepended to the list of elements processed before it, resulting in the list being reversed. So, y = [4,3,2,1]. In the other case, the same operation occurs from right to left, resulting in the list being reconstructed in its original order, with z = [1,2,3,4].

7.6   Exercises

1. A common task in problems that involve graphs is the determination of the degree of the vertices in the graph. Assume that we have a graph represented as an adjacency matrix, where $A_{ij} = 1$ means that an edge from vertex i to vertex j is in the graph, and 0 if no such edge exists. For an undirected graph where $A_{ij} = A_{ji}$, we can perform this computation of degree using matrix-vector multiplication. Describe how you would express this using pseudocode. Assume that you have an intrinsic function SUM() available that computes the sum of elements in an array passed to it as an argument.

2. How does the Actor model prevent certain concurrency control issues? How does this change how a programmer might implement a program?

3. Write a program with a race condition using a two-sided message passing notation of your choice (e.g., MPI). How easy would it be to accidentally introduce a race in a real program?

4. One of the most important numerical kernels is the multiplication of a sparse matrix (mostly zeros) with a dense vector. This is the basis of most iterative solvers. Describe how the sparse matrix vector multiplication problem can be expressed in terms of a sequence of reductions.

5. Is a reduction operation over a set of floating point numbers deterministic?

6. Is a reduction operation over a set of integers deterministic?

7. Physics simulation codes (used in scientific computing but increasingly in computer games as well) often need to solve a partial differential equation. In these problems, a grid is superimposed over a problem domain and points are updated based on values of neighboring points (and explicit finite difference solver). These usually appear inside a loop to model how a system evolves over time. Without getting bogged down in specific details, discuss how effective functional languages would be for these types of problems.

8. Discuss the relative strengths and weaknesses of occam, OpenMP, and functional languages using the cognitive dimensions from Chapter 5.

Chapter 8

Performance Considerations and Modern Systems

Objectives:

- Discuss the issue of CPU versus memory performance and the impact this difference has on program performance.
- Introduce Amdahl's law and the concepts of parallel program speedup and efficiency.
- Discuss the performance considerations and potential consequences of lock-based synchronization.
- Introduce the concept of performance overhead when using concurrency constructs.

In this chapter, we will explore the important issue of performance. Ultimately, concurrency is most often utilized to increase performance. One moves to a parallel context when it provides a performance benefit not present when using traditional sequential methods. As performance is the motivating factor leading to the use of concurrency, it is important to understand the performance properties of concurrent systems. The primary factors that affect program performance include:

1. Hardware performance.

2. Synchronization and coordination control.

3. Operating system and runtime overhead.

In sequential programming, we are taught and learn through practice that there are specific ways we should structure programs to perform well given the performance characteristics of a typical computer. For example, we are taught to write code that exploits locality if we wish to benefit from the memory efficiencies provided by cache-based memory hierarchies. We also learn that sometimes managing resources manually within a program is preferable to allowing the operating system to do so and suffering the high cost of system calls. Over the career of a working programmer, experience with different projects and platforms leads to an accumulation of knowledge and wisdom about how to achieve good performance in different programming scenarios.

While this performance knowledge and folklore that we have built up regarding sequential systems still applies to concurrent programs, a new set of performance considerations exists that are specifically related to concurrency. For example, modern multicore processors continue to take advantage of caches and multi-level memory hierarchies. From a performance perspective we still are concerned with locality, but now must add the complication that arises when multiple cores with their own caches share a single common memory. A programmer who has worked with sequential programs only may not appreciate the impact this might have on the efficiency of programs and their use of memory, especially given the requirement that the cache must maintain a consistent view of memory for all of the cores.

In discussing performance, we will point out how features of the languages so far relate to these issues. After all, this text is focused on language-level topics, not hardware architectures or operating systems. Unfortunately, the topics are not mutually exclusive. To illustrate this, we will discuss language features for concurrency that can actually yield very poor performance when used improperly due to effects outside of the scope of the language. Similarly, we will show how the use of higher-level language constructs allows automated tools to make decisions related to performance that would be very difficult to implement by hand with lower-level techniques.

8.1 Memory

Memory is one of the most common sources of performance problems in programs, both sequential and parallel. Inefficient use of memory can lead to performance degradation. In sequential programs, performance often suffers if a program traverses memory in an irregular or cache-unfriendly manner resulting in most of the execution time of the program being spent waiting for data to move between the processor and parts of the memory hierarchy. The reason memory is the source of performance degradation is the significant gap between the time for a single CPU cycle and the latency time for memory accesses. In fact, this growing gap between the cycle time in the processor and the time required to access memory has been a primary motivation behind architectural features like caches, out-of-order execution logic,

TABLE 8.1: Overheads associated with common activities in a typical SMP system.

Operation	Minimum overhead (cycles)	Scalability
Hit L1 cache	1-10	Constant
Function call	10-20	Constant
Integer divide	50-100	Constant
Miss all caches	100-300	Constant
Lock acquisition	100-300	Depends on contention
Barrier	200-500	Log, linear
Thread creation	500-1000	Linear

and superscalar pipelined processors. This gap has been growing for most of the time computers have existed, and was termed the *von Neumann bottleneck* by John Backus in 1977 [8] due to the computing model named for John von Neumann that most computers are based upon.

Of the many architectural features proposed and implemented to address the processor/memory gap, caches are of particular interest when it comes to their role in concurrent systems. Other features (such as pipelining) have less of an impact on concurrent programs, primarily because they are focused on the execution of sequential threads of execution within a single processor core. Interactions between these sequential threads of execution where coordination errors and performance issues can arise is typically achieved through the exchange of data through a memory. Therefore the cache and components of the memory hierarchy lie directly on the critical path that concurrency coordination operations take, and have an significant potential impact upon them. We will start with a brief refresher on single-processor caches before discussing them in a parallel context.

8.1.1 Architectural solutions to the performance problem

Historically, there have been a variety of architectural features introduced to overcome this performance gap to hide memory latency from programs. The most widespread technique that all readers should be familiar with is the multilevel memory hierarchy , where *caches* play a critical role. In cache-based systems, small memories are provided that sit between the processor and main memory with significantly lower latencies and smaller capacities than the main store. For example, in a single layered cache-based architecture, between the processor and main memory would lie a small (relative to main memory) block of memory that can be service requests from the processor much faster than main memory could. The goal of the cache is to replicate a subset of main memory in this fast memory closer to the processor in hopes that accesses from the processor will be satisfied by the cached copy instead of the slower main memory. In Table 8.1, we show typical overheads associated with common

Listing 8.1: Code with poor cache performance.

```
real, dimension(8000,8000) :: X
integer :: i,j

do i=1,8000
   do j=1,8000
      X(i,j) = 42.0
   end do
end do
end
```

activities, including cache misses, in a typical shared memory multiprocessor.

Every time a memory access is issued, the cache first checks to see if it contains the data at the address being accessed. If it does, it can respond with the data very quickly. If it does not, the memory system retrieves the relevant data from main memory, storing it *and* its neighbors (data at nearby addresses) in the cache before returning the requested data to the processor. If subsequent accesses from the processor are made on addresses near the one associated with the request that went out to main memory, the fact that the memory subsystem retrieved data nearby the original request and stored it in the cache means that it is likely these subsequent accesses could be fulfilled by the cached data. The result is higher performance by amortizing the high cost of retrieving blocks of data from main memory with the high likelihood that sequences of memory requests would be nearby each other in terms of addresses.

Unfortunately, the cache-based approach is not perfect. Not all programs work by addressing memory linearly in a very regular pattern. Programs often stride though memory in regular steps that are larger than the spatial window that is provided by the cache. In other cases a code may simply perform unstructured access patterns where no regularity is present and the occurrence of locality would only be due to chance. Multi-level caches are able to deal with this if the striding or irregularity of the access pattern is constrained to relatively small regions of the address space.

8.1.2   Examining single threaded memory performance

To illustrate how single threaded programs can experience performance problems due to improper use of the memory hierarchy, consider the basic algorithm in Listings 8.1 and 8.2.

In both cases, we have a two-dimensional array over which we iterate and set all elements to some constant value. The only difference between the two versions is the order of the loops and correspondingly, the rate at which

Listing 8.2: Code with good cache performance.

```
real , dimension (8000 ,8000)  ::  X
integer  ::  i,j

do  j =1 ,8000
    do  i =1 ,8000
        X(i,j)  =  42.0
    end do
end do
end
```

the first and second index change as the loop iterates. This choice of array traversal pattern is related to the choice the underlying language makes about how arrays are laid out in memory.

Row versus column-major ordering

A very common source of poor memory traversal resulting in bad cache utilization is related to multidimensional array layout. In languages like C, a *row-major* ordering is used. This means that for a two-dimensional array, the rows of the array are stored consecutively in memory. The first element of the array corresponds to the first element of the first row, and the second element corresponds to the second element of the first row. On the other hand, *column-major* languages store columns consecutively in memory. In this case, the second element of a two-dimensional array would be the second element of the first column, not the second element of the first row. Fortran is an example of a column-major language, as is the popular MATLAB® package. This ordering difference extends beyond two-dimensional arrays to those of higher dimension.

Often people talk about this ordering problem in terms of which index changes fastest. Consider a two-dimensional array with m rows and n columns. In C, as one traverses memory in order, the column number increments (modulo n) with every address while the row increments every n accesses. Fortran on the other hand increments the row with each consecutive address (modulo m) with the column incrementing every m accesses. So, in C the column varies fastest while the row is the fast dimension in Fortran.

When translating algorithms between languages (or, from pseudocode to implementation), it is easy to create an algorithm that is perfectly correct but performs poorly simply due to an inappropriate ordering of loops. Consider the example in Listing 8.1. In this case written in Fortran, the algorithm varies the column number fastest. This means each iteration accesses data n elements away from the last. If n is sufficiently large, each iteration will request memory that is outside the bounds carried into the cache by the previous iteration, thus requiring a costly request to main memory. The per-iteration time will be very high, leading to a slow algorithm. On the other hand, simply swapping the loops will result in the identical output (in this case), but consecutive accesses will be contiguous in memory, allowing the slow main memory accesses to be absorbed by the higher performing cache accesses, resulting in an overall performance improvement.

We show this example of single threaded memory performance and caching issues to illustrate the concept that the underlying hardware can have effects on the performance of program code without impacting its correctness. Understanding this simple case with a single threaded program is important in order to understand similar issues that arise with multithreaded programs that are implemented on a set of processing elements with their own separate caches and a shared main memory. Multicore systems are an example of this hardware design that we see in practice.

The problems posed by the two array orderings can also impact codes written in multiple languages where arrays are shared amongst the languages. For example, a popular way of writing extensions for MATLAB is to utilize the C API that the package provides. When an array is passed to C from MATLAB, a C programmer must be conscious that while arrays allocated by C use a row-major format, those allocated by MATLAB and passed to C are laid out using column-major. C programmers must restructure their loops accordingly to avoid problems with cache performance. If the C code uses multiple threads, the performance impact of not recognizing the column-major format of the MATLAB allocated data can be even more damaging to overall performance. We will see why in the next section.

8.1.3   Shared memory and cache coherence

In a shared memory system with a cache-based memory hierarchy, the system must ensure that each processing element sees a coherent view of the main memory. The basic protocol that a single processor cache hierarchy implements to guarantee memory integrity is summarized as ensuring that any modifications made to data that are stored in cache are written safely to main memory before they are replaced by other data during a cache miss. For cache hierarchies with more than one level, this translates to writing to a level of cache closer to main memory when a miss occurs, which is conceptually no different. This is sufficient for a single processor system, but what about a case where more than one processor (with their own independent caches) shares a single main memory?

The issue we are faced with is based on the assumption that, should one processor modify memory, the modification should be seen by all other processors that read the modified memory. Maintaining this coherent view of main memory is complicated by the presence of caches. For example, a processor may modify a value but only maintain the modified value in its local cache because it may have not yet performed an operation that requires the cached

update to be flushed back to main memory. A coherent view of memory requires that, should another processor attempt to read that modified memory
before the cached value on the modifying processor is committed to memory,
the reader sees the most recent value instead of the potentially out-of-date
value that remains in main memory. To deal with this, shared memory parallel architectures utilize what is known as a *cache coherence protocol* to ensure
that all processors see a consistent view of memory even if processors have
not yet forced modifications to be committed from cache to main memory.
Multicore processors utilize these cache coherence protocols to maintain this
consistent view of memory amongst their set of cores.

The details by which cache coherence protocols function is beyond the scope
of this text, and is discussed in great detail in books on parallel processor architecture such as Culler and Singh [25]. It should suffice to state that cache
coherence protocols require some level of overhead when data from the main
memory is accessed that has been modified and stored in a cache of a single processor. The reason that we discuss cache coherence here is because
the presence of a cache coherence protocol can have an adverse performance
effect that is not apparent from simple examination of code. Furthermore,
the effect of ill-use of a cache coherent system is often observed only by performance degradation and not correctness. The beauty of a cache coherent
system is that programs run as expected, with much of the magic required
to provide a coherent view of memory hidden from the programmer and left
to hardware-level protocols. Poor memory usage can be difficult to pinpoint
because programs run correctly, but sub-optimally only with respect to their
runtime performance. We will demonstrate this with a simple example.

Memory traversal and coherence protocol interaction

Consider a program that has the simple task of taking a very 1D large array
and iteratively updating the elements of the array to contain the average of
each value and its neighbors. Mathematically speaking (ignoring boundary
elements), this is simply the act of computing

$$x_i = \frac{x_{i-1} + x_i + x_{i+1}}{3}.$$

This can be computed by P processing elements in many different ways,
two of the most extreme being illustrated here. In one case, we split the
array x containing N elements into $\frac{N}{P}$ element contiguous chunks, with each
processor iterating over each sequence element by element. The C code for
a thread body performing this work is shown in Listing 8.3. In a second
case, we decide that each processor strides through the array by steps of P
elements ranging from 1 to N. The C code for this thread body is shown in
Listing 8.4. The key difference is in the start, end, and stride values used
for the inner loop (these variables are italicized in the listings). Note that both
example listings include code for double buffering and thread synchronization

Bad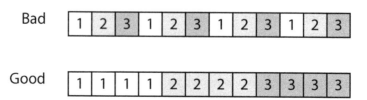

Good

FIGURE 8.1:   The difference between a good (spatially local) and bad (strided) access pattern for three threads working on a shared 1D array.

to guarantee deterministic behavior when these workers are run in multiple threads sharing the `data` array.

What we observe in terms of performance is quite striking, although quite obvious after close examination of the code and a recognition that a coherence protocol exists in the memory system. In the first case where the data is split into coarse chunks upon which each processor iterates over individually (Listing 8.3), we see that each processor reads from locations that it is the sole modifier of for all but the boundary cases at the beginning and end of the chunks they are assigned.

On the other hand, the second case (Listing 8.4) exhibits the property where each computation on an element requires reading a value for the immediate successor and predecessor that were modified by different processing elements. These successor and predecessor values are likely to exist in the caches of the respective elements that modified them. The local cache of each processor is therefore largely useless, as the neighbor values that were read into the cache from main memory are not likely to contain the modified values which most likely lie in the caches of other processors. The result is that each memory access will require either an access to a remote cache containing the current element value, or a flush to main memory and subsequent read. The precise set of operations that occur to provide this coherent view of memory is dependent on the specific cache coherence protocol that the system implements. We illustrate a the access patterns exhibited by these two examples in Figure 8.1, in which the 12 element array is labeled by the thread that computed and wrote each element.

As we saw earlier, the traversal pattern of a single threaded program can suffer performance degradation due to poor locality in its memory usage. This example demonstrates that when we are faced with a cache-coherent shared memory, we can suffer further degradation due to the interactions between the underlying caches that provide each processor a coherent view of memory.

This leads us to an interesting performance issue. As we move higher in abstraction levels, we yield fine grained control over how the computer implements an algorithm to the language compiler and runtime system. By handing off concerns such as the order of array traversal for a whole-array operation,

Listing 8.3: A thread body that computes over contiguous blocks of the shared array.

```
static double *data, *dTmp;
static int N, P;

void worker_friendly(void *arg) {
  int rank = (int)arg;
  int start = rank*(N/P);
  int end = (rank+1)*(N/P);
  int stride = 1;
  int i,iter;
  int numiters = 256;
  double *tmp, *dcur, *dprev;

  /* use double buffering to ensure deterministic
     computation of loop in the presence of multiple
     threads */
  dprev = data;
  dcur = dTmp;

  for (iter = 0; iter < numiters; iter++) {
    for (i = start; i < end; i += stride) {
      if (i > 0 && i < N-1)
        dcur[i] =
            (dprev[i-1]+dprev[i]+dprev[i+1])/3.0;
    }

    /* sync to make sure all threads are at this
       point before swapping buffer pointers for
       next iteration */
    sync_threads();

    /* swap pointers */
    tmp = dcur;
    dcur = dprev;
    dprev = tmp;
  }

  /* copy dTmp into data if dcur pointed at
     dTmp in final iteration */
  if (dcur == dTmp)
    memcpy(data, dTmp, N*sizeof(double));
}
```

Listing 8.4: A thread body that computes over a strided set of elements in the shared array.

```
static double *data, *dTmp;
static int N, P;

void worker_unfriendly(void *arg) {
  int rank = (int)arg;
  int start = rank;
  int end = N;
  int stride = P;
  int i,iter;
  int numiters = 256;
  double *tmp, *dcur, *dprev;

  /* use double buffering to ensure deterministic
     computation of loop in the presence of multiple
     threads */
  dprev = data;
  dcur = dTmp;

  for (iter = 0; iter < numiters; iter++) {
    for (i = start; i < end; i += stride) {
      if (i > 0 && i < N-1)
        dcur[i] =
          (dprev[i-1]+dprev[i]+dprev[i+1])/3.0;
    }

    /* sync to make sure all threads are at this
       point before swapping buffer pointers for
       next iteration */
    sync_threads();

    /* swap pointers */
    tmp = dcur;
    dcur = dprev;
    dprev = tmp;
  }

  /* copy dTmp into data if dcur pointed at
     dTmp in final iteration */
  if (dcur == dTmp)
    memcpy(data, dTmp, N*sizeof(double));
}
```

TABLE 8.2: Latencies observed in an Itanium-2 based distributed memory system with both Infiniband and Gigabit Ethernet interconnection networks.

Source of latency	Cycle count
L1 cache	1-2
L2 cache	5-7
L3 cache	12-21
Main memory	180-225
Gigabit Ethernet (access to remote node)	approx. 45,000 (30 μs)
Infiniband (access to remote node)	approx. 7,500 (5 μs)

or data layout of a complex data structure in memory to a compiler, we can allow it to determine the ideal implementation based on its knowledge of the target platform. On the other hand though, we lose the ability to perform tuning when the high-level abstraction is mapped poorly onto the hardware for specific applications. Often the best approach is to opt for high-level abstractions, and drop to lower levels when performance is critical and it is clear from performance measurements that the compiler is not generating optimal code.

8.1.4 Distributed memory as a deeper memory hierarchy

In some concurrent systems, we do not have a single shared memory that all processing elements can see, but a set of disjoint memories owned by subsets of processors that can be treated as a single, *distributed memory*. The key feature of a distributed memory architecture is that memory is not addressable at the instruction set level between some pairs of processors. For example, in a cluster of PCs, individual computers cannot address memory on others without explicitly communicating with them and bringing copies of the remote values into their local memory to perform work on. This is where message passing plays a role in building parallel programs for these types of machines, as the exchange of messages over a network allows processors to access data in memories other than their own.

From an abstraction perspective, one can quickly see that distributed memory architectures are simply creating a new layer of memory hierarchy with a latency of access defined by the interconnection network. Local main memories at each processor act as caches above the larger distributed memory space. If a processor requires data that it does not hold in its local memory, it must make a request to the owner of the memory holding the data to gain access to it. The requester can store this data locally in its memory while it works on it, and must be sure to provide it back to the original source if it is necessary for others to be able to see any modifications that it makes. In Table 8.2 we see some typical latencies both within a single node and across nodes for a cluster based on Itanium 2 processors with Gigabit Ethernet and Infiniband interconnection networks.

Unfortunately, providing coherence in the distributed memory context is rarely assisted in hardware and relies upon a software layer to provide the equivalent of a shared memory semantics on top of the physically distributed memory. Software layers that provide this *distributed shared memory* have been proposed and developed over many years, although their acceptance is limited. The key limiting factor is that shared memory cache coherence can be implemented very efficiently using hardware techniques such as bus snooping to allow caches to be synchronized simply by watching signals that processors send over the bus to the main memory. Interconnection network technologies used in common distributed systems rarely provide this functionality in an efficient manner, so coherence protocols built on distributed architectures tend to be limited in their performance.

8.2   Amdahl's law, speedup, and efficiency

In an ideal world, all parallel programs would scale perfectly and the application of p processors to a problem would yield a speedup of p. This would mean that the time for a parallel program to execute, t_p, would be a fraction of the original sequential version.

$$t_p = \frac{t_1}{p} \tag{8.1}$$

A parallel program taking time t_p from Equation 8.1 is, unfortunately, highly unlikely to be achieved in any practical application. There is a simple reason for this. It is hard (often impossible) to avoid having some part of a program require sequential execution within a single thread of control. The reasons for this are very application specific, although one can consider a simple case that is present in most programs. We will see later that locking and other coordination control schemes contribute to this serialization of parallel programs.

Most programs that execute in parallel have some `main()` procedure that sets up the input for the portions of the program that execute concurrently. If we assume that the concurrent portion of the program is perfectly parallelizable (often informally referred to as "embarrassingly parallel"), the time that it requires to execute follows Eq. 8.1. Clearly as p grows larger, t_p grows smaller — in the limit of ever increasing parallel execution unit counts, it approaches zero.[1] Unfortunately, there is still the sequential setup cost that is not impacted in a positive way by adding parallel processing capabilities. Assuming that the time for this setup is not a function of p, then this sequential time t_s is fixed. The overall runtime of the program on p processors is therefore $t(p) = t_s + t_p$, which in the limit, is t_s. The end result is that the program can never get faster than the sum of its components that are inherently sequential.

The speedup of a program that has *any* components that are not concurrent can never be perfect as it is ultimately bound by the intrinsically sequential parts of the program. This fact is known as *Amdahl's law*. In practice, the intrinsically sequential parts of a program extend beyond simple setup and finalization stages in the main body of the program. Often there are sequential operations that occur when threads synchronize between regions of parallel computation. In fact, aggressive application of locking and mutual exclusion schemes to protect critical data can serialize a parallel program, resulting in significant performance degradation.

Many people come to concurrent programming expecting huge, almost miraculous speedups, only to be ultimately brought back to reality by Amdahl's law. Surprisingly, we all have experience with Amdahl's law in everyday life and make plans based on it — we simply often haven't worried about it as a formal, quantitative issue. Consider the problem of shopping at a store during a big holiday sale. Assuming the store isn't too crowded, we would expect to browse around and select our goods to purchase at the same rate as we would if we were alone in the store. From the perspective of the store, the increase in customer count corresponds to a higher rate at which goods leave the shelf.

To the store, goods are like computations to be performed, and the consumers are simply processors that are picking up these waiting tasks. Unfortunately, the number of checkout lines is fixed, so regardless of the number of customers in the store, the maximum rate of checkout does not increase during the holidays. If we consider the goods to be computations, then having a customer count equal to the number of goods corresponds to all of them being simultaneously "computed." Unfortunately, the rate at which customers can check out is limited. So, at some point there is no point in adding additional customers, as they will simply queue up and wait to check out, and the throughput of goods being purchased will stop increasing.

---

[1] Obviously t_p cannot actually be zero, as the concurrently executing parts of the program must execute at least one instruction. This would require at least one clock cycle.

8.3   Locking

The main concern of a programmer targeting a parallel platform is correctness first, with performance secondary. It is of little use to a user of a program to rapidly compute an incorrect result. Correctness and predictable behavior are critical. As we saw earlier, to ensure correctness in the presence of concurrency, one must provide mechanisms that prevent nondeterministic behavior when it can result in inconsistent or unpredictable results. This requires the programmer to protect data within critical sections and implement complex operation sequences as though they were atomic. This allows the programmer to control non-determinism and avoid pathological conditions such as races and deadlock. The basic primitive to implement this coordination between concurrently executing threads is the lock. Unfortunately, as most programmers experienced with threading and other forms of concurrency are aware, locks come with a price.

As we just discussed in the context of Amdahl's law, some portion of most concurrent programs is intrinsically sequential. The overhead that can come from locks that adversely impacts performance results from:

- Serialization of threads at critical sections.

- Wasted time waiting for lock acquisition.

- Time spent acquiring and releasing locks to protect data and code rarely accessed by more than one thread at a time.

8.3.1   Serialization

By definition, a critical section is a region of code that must be entered by one and only one thread at a time to prevent correctness problems in a concurrent program. If more than one thread encounters a critical section at the same time, the semantics of the critical section ensure that one and only one process executes it at a given time. This means that one thread makes progress while others wait for permission to enter the section. In the limiting case, the entire program is a critical section. If the program takes t time to execute, then p threads will take tp time to execute, as only one may execute at a time. Obviously the granularity of critical sections is most often smaller than the entire program, but if a critical section is a piece of code that is executed very frequently by all threads, then the likelihood of serialization instead of parallel execution increases.

The concept of a transaction was invented to reduce the amount of serialization that results from the presence of code sequences that require atomic execution. Mutual exclusion with locking forces single threaded access to the critical section and the data it works with. A transaction on the other hand allows access to data that a critical section works with by multiple threads, and provides a disciplined method for identifying and resolving conflicts that can arise when threads use this sensitive, shared data in a dangerous way. This allows for some level of concurrency within the critical section, with the restriction of mutually exclusive access isolated in a phase where the results of the execution of the critical section are verified and committed to memory at the very end.

When the sequence of operations that act on this data is much larger than the set of operations necessary to check and commit it, we can see reduced serialization in the code and higher degrees of concurrency that can be achieved. In the context of Amdahl's law, this means that transactions help us reduce the amount of time our program spends in a sequential mode of execution, increasing the achievable speedup due to parallelism.

8.3.2   Blocking

Inevitably, a concurrent program containing locks will have at least one thread attempt to acquire a lock that another thread already holds. In this case, the thread that is attempting to acquire the lock will be unable to proceed until the thread that currently holds the lock releases it. In a conservative locking scheme, a program will acquire locks for all of the data elements that it will work on within a critical section before entering the section. The result will be that other threads may attempt to access this data and be unable to acquire the lock even if the thread that holds the lock has not actually operated on the data yet. Even worse, the thread may never operate on the locked data if the operation on this data is protected by a conditional operation that does not execute.

We can alleviate this by adopting a two-phase locking scheme within a critical section. In a two-phase locking scheme, locks are acquired immediately before the data they protect is accessed. This allows a thread to enter a critical section without locking all of the data within the region until it actually requires it. Therefore, other threads may be able to access this data while it is still "untouched" by the critical section that one thread is executing. This phase of acquiring locks as data is accessed is known as the *growing* phase, as the set of locks held by a thread grows during this period. When the critical section completes, the thread holding the locks releases them. No lock is released in the critical section until it completes. This phase of lock release is called the *shrinking* phase.

Why is two-phase locking useful? Remember that a critical section represents a potentially large sequence of operations that must execute atomically. A consequence of atomicity is that any intermediate value computed by this sequence is not visible to other threads — that would violate the atomic nature of the sequence. In two-phase locking, locks are acquired only when the critical section requires them. So, another thread may access one of these

Listing 8.5: Illustration of unnecessary whole-method locking.

```
public synchronized void f() {
    // safe code ...

    if (rare_conditon == true) {
        // operations requiring concurrency control
    }

    // more safe code...
}
```

values while the critical section is executing in a different thread, but before it actually requires it. This other thread will be able to access the value then, avoiding the blocking scenario that would result if a critical section acquired all locks before starting. The result of this is a lower probability of two threads serializing.

Two-phase locking and other similar schemes of relaxed lock acquisition do not eliminate potential for deadlock, but they do increase the potential for concurrency by eliminating potential situations where blocking on lock acquisition may occur. The idea of two-phase or other incremental lock acquisition schemes comes from the concept of a transaction. At the language level, software transactional memory schemes can aid in avoiding these unnecessary blocking scenarios if the underlying STM system uses these locking schemes.

8.3.3 Wasted operations

Locks, when used correctly, protect the program from *potential* correctness problems. The key word here is potential. In some cases, the sequence of events leading to incorrect execution without the locks is rare — code without locks would work "most of the time."[2] Unfortunately, rare doesn't imply impossible, so circumstances can exist that will result in correctness problems on the off chance that one of these improbable executions occurs. So, a programmer that wishes to build a program that never falls victim to these improbable (but possible) situations will ensure that locks for data are properly acquired to prevent problems. The consequence of this is that the program will spend some amount of time acquiring the lock and releasing it even if the situation that it is preventing doesn't occur. For example, consider the Java subroutine in Listing 8.5.

---

[2]We place this in quotes because, while programmers often make this statement, they rarely can quantify the actual frequency that a user will encounter an erroneous execution.

The application of the synchronized keyword and subsequent locking of the entire method call will cause other threads accessing the object containing $f()$ to block. This is true even if the condition that executes the sensitive region of code within the if-statement is very rarely encountered. What is worse, programmers may be more conservative than they need to be, and may impose a locking discipline to protect data from circumstances that will *never* happen. In sufficiently complex programs, this is a case that is sometimes difficult to distinguish. This is not uncommon when programmers do not fully understand the aspects of their programs that require concurrency control relative to those parts that do not.

8.4   Thread overhead

Threads are not free. Whenever a program wishes to branch its control flow into two or more threads of execution, some overhead exists for starting the additional threads necessary to perform the parallel computation. In the worst case, a thread must be created from scratch, often requiring some interaction with the operating system or runtime layer to allocate resources and perform scheduling. Similar overhead may be required when the thread completes to properly deallocate it from the system. Techniques such as thread-pooling attempt to avoid this by pre-allocating the execution contexts for each thread once, and keeping the threads around in an idle state until they are necessary. The thread contexts are also reused in order to avoid the overhead of destroying a thread that can instead be recycled for other work later on. Even so, reactivation of an idle thread from a pool requires some degree of overhead to allow the operating system or runtime scheduler to recognize that it has become active and must be scheduled onto available compute resources.

This overhead for thread creation, activation, and destruction must be taken into account when designing parallel algorithms to perform well. The reason is that the speedup realized by a program is limited by this unavoidable overhead. In fact, overhead in the underlying system unrelated to the algorithm as programmed by the programmer is often the reason why "perfect speedup" is unobtainable for nearly all parallel algorithms. Management of threads often is implemented using critical sections in the operating system or runtime, so a program that itself is parallel may experience serialization as it interacts with this underlying support layer. This serialization means that Amdahl's law applies, limiting the ultimate speedup of the code.

Consider a unit of computation that a programmer decides should be split up and executed in parallel. Assume that the original computation takes T units of time, and the parallelized version on p processing elements is com-

Listing 8.6: A naive multithreaded mergesort.

```
array mergesort(array X) {
   int len = length(X);
   int half = length(X)/2;

   if (len <= 1) return X;
   else {
      array left = spawn mergesort(X[0..half]);
      array right = spawn mergesort(X[half+1..len]);
      wait(left);wait(right);
      return merge(left,right);
   }
}
```

posed of p blocks of work that take $\frac{T}{p}$ units of time to complete each. The actual time to execute each block of work on the parallel processing elements will actually be

$$T_{block} = T_{startup} + \frac{T}{p} + T_{completion}.$$

If the startup and completion overheads for the threads are essentially zero relative to $\frac{T}{p}$, the speedup will be near-ideal. Unfortunately, if the granularity of the work performed in the parallel blocks is very small, programmers could find that their speedup is far less than expected and dominated by system overhead outside of their program.

This situation is surprisingly easy to encounter in what appears to be the most innocuous of algorithms.

To demonstrate this problem, we will consider the traditional mergesort algorithm. We will implement it in the same manner that it is implemented in its traditional sequential version with a simple extension. Instead of recursively calling the algorithm through standard recursion, we will instead spawn a new thread of execution for each recursive call with the caller blocking until each child thread completes before performing the merge step. This is shown in Listing 8.6, where we ignore the merge step since it is assumed to be the standard sequential **merge()** routine that the reader should already be familiar with.

A programmer may be inclined to try this as a first shot at parallelizing the sort algorithm for very large inputs because, in the early stages of the algorithm the number of elements in each half is also very large. In fact, the author when learning some of the languages discussed in this text (in this case, the Cilk language based on C) did this very thing. What is quickly realized when executing the code is that the parallel version is quite likely going to run very slowly, much slower than expected. Why does this occur? The answer is simple, but subtle.

Consider the recursion tree that is executed by the algorithm. From basic discrete mathematics, we know that for a tree containing $2^N - 1$ nodes, 2^{N-1} of these nodes are leaves. In the case of the sorting algorithm, leaf nodes correspond with the case where 1 or zero elements are sorted — or more bluntly, leaf nodes correspond with nothing interesting happening algorithmically than simply returning that which was passed into the function call. In these cases, the time spent performing a computation is negligible — nothing really happens inside the call. So, the time for each of these 2^{N-1} calls is dominated by the overhead to make the function call. In the multithreaded case, the time is therefore dominated by the overhead to create and destroy threads instead of actually performing useful work. For any non-trivial array of length N, the value of $2^{N-1} * (T_{startup} + T_{completion})$ is something that cannot be ignored. In fact, it will dominate the runtime of the parallel algorithm leading to abysmal speedups. If programmers encounter this and do not dig into the root of the performance problem, it is entirely possible that they will disregard the value of multithreading altogether!

The solution to this problem is to identify the point where the time for the sequential version of the mergesort algorithm is large enough such that it dominates the thread overhead, but small enough such that further divisions of the work will begin to see their time overwhelmed by overhead. This is both platform and problem dependent, and how to identify this point is left to the later chapters where we discuss the recursive decomposition programming pattern in more detail (Chapter 12). For now, assume that we have identified an array length N_τ that we consider to be the largest length where the sequential version of the algorithm beats the parallel implementation due to overhead in thread management. We can therefore refine the algorithm from Listing 8.6 to take this into account, yielding the version in Listing 8.7 in which we can expect a reasonable benefit to performance from parallelism.

Performance is an art learned through practice. In the case of thread overhead, the art is understanding when there is a benefit to spawning a new thread of execution versus executing a traditional sequential algorithm. The overhead for threading and the time requirement for a unit of work is, unfortunately, tied intimately with the platform on which the code will execute. Thus, deciding when to go parallel versus staying sequential is a tricky yet important aspect of designing parallel algorithms.

Listing 8.7: A smarter multithreaded mergesort.

```
array mergesort(array X, int Nt) {
   int len = length(X);
   int half = length(X)/2;

   if (len <= 1) return X;
   else {
     array left,right;
     if (len > Nt) {
       left = spawn mergesort(X[0..half]);
       right = spawn mergesort(X[half+1..len]);
       wait(left);wait(right);
     } else {
       left = mergesort(X[0..half]);
       right = mergesort(X[half+1..len]);
     }
     return merge(left,right);
   }
}
```

8.5   Exercises

1. Write a program that allocates a very large (>100MB) array of integers and measures the amount of time to read and write from random locations within increasingly large regions of the array.

   For example, let $N = 5,000,000$ be the number of times the operation `a[X] = a[Y]+1` is executed, where X and Y are random indices drawn from the increasingly large ranges of the array a. We could measure the amount of time to perform N operations over the first 8, 16, 32, 64, and so on, elements of a.

   Plot the times to perform N operations for each of these range sizes and explain what you see with respect to the size of the various caches on your system. Be sure not to create an array that is too close to the RAM capacity of your machine, as you may observe effects of swapping instead of the caches alone.

2. Write a program in C that uses a two-dimensional array of floating point numbers with each dimension of the array consisting of 1000 or more indices. The program should execute a doubly-nested loop (one for each index of the array) where each element is updated with the average of

its four neighbors (up, down, left, right), where the boundaries wrap around (e.g., the upper neighbor of array elements on the top row are the elements in the same column on the bottom row).

Measure the amount of time required to execute a number of iterations of the doubly nested loop. Now, switch the indices that each loop corresponds to and measure how the timing of the loop changes. Discuss what you see in relation to row-versus-column major ordering and caches.

3. Using the example of cooking a meal in the introduction, identify the parts of the process of preparing the meal that are inherently sequential and those that can be parallelized. How does Amdahl's law impact show up in real-world activities?

4. You have a program that is composed of an initialization step that must run sequentially on a single processor taking $t_1(n) = T_1$ time, where n is the input size for the program. After initialization, the program executes a loop that can be parallelized perfectly which requires $t_2(n) = T_2$ to execute on a single processor. This is followed by a shutdown phase that executes in parallel but requires $t_3(n) = T_3 \log(p) + 1$ time, where p is the number of processors that are used.

Assume that T_1, T_2, and T_3 are all the same. Plot how the total time $t(n) = t_1(n) + t_2(n) + t_3(n)$ changes as a function of p. Write $t(n)$ in terms of T_1, T_2, and T_3. When p becomes large, which term of $t(n)$ dominates?

5. Consider a program that is sequential when it starts, eventually spawning a set of threads to perform an activity in parallel, and then resuming sequential execution after the threads complete. To tune the program, we may wish to identify the time difference between the earliest and latest threads to start and the earliest and latest to complete. These time differences will allow us to identify time that some processors waste sitting idle that we can eliminate to increase machine utilization. Assume that the threads all have access to a common timer. Using only four variables corresponding to the first thread to start, last to start, first to finish, and last to finish, show how these times can be recorded without using locks to protect these variables. Assume that threads can write to the variables atomically.

Chapter 9

Introduction to Parallel Algorithms

Objectives:

- Discuss approaches to identifying and exploiting concurrency in programs.
- Introduce algorithmic and source code patterns for concurrent programming.

So far we have explored the basic topics any programmer wishing to work with concurrency must understand. These topics include:

- The basic correctness and performance considerations that arise in concurrent programming. These are vital for designing correct programs that will perform well.

- The basic properties of common sequential programming languages to build an understanding of what high-level languages provide, and most importantly, the barriers that these languages put up when applied to concurrent applications.

- Language constructs specifically designed for concurrency and parallelism over the history of language development, both with respect to their origins and the properties of their concurrency constructs.

In the remainder of the text, we turn our focus to how language constructs added specifically to support concurrency interact with the design and programming of parallel algorithms. Hence we need to pause in our discussion of high-level languages and spend some time considering parallel algorithms and their design. The reader should note that we will frequently use the term *parallel* interchangeably with *concurrent* for the remainder of the book. This is intended to reflect the terminology most commonly used in the literature for these topics. As discussed at the beginning of the text, the two terms are intimately related.

9.1   Designing parallel algorithms

It is easy to be overwhelmed when first approaching the topic of parallel algorithms. All the challenges of designing a serial algorithm remain, but in addition parallel programmers must decompose their problem into smaller subproblems that can run at the same time, manage data to avoid memory access conflicts, and control data movement to maximize performance.

Fortunately, the field of parallel programming is many decades old. We've learned a great deal about parallel algorithms and how to "think parallel." The challenge is how to concisely share this body of knowledge with programmers new to parallel programming.

The object-oriented programming community faced a similar problem in the early 1990s. In a now famous book [37], four authors (now known as the "gang of four" or just "GOF") wrote down a catalog of key design patterns encountered in object-oriented programming. A *design pattern* is a well known solution to a recurring problem within a well defined context. They are written down in a consistent format so a programmer can quickly understand the problems addressed by the pattern and how to use the pattern in solving a new problem. The GOF book revolutionized how people learned and talked about object-oriented programming. These patterns established a new terminology that designers could use when talking about object-oriented design. Essentially, the book helped the field of object-oriented programming grow up and become the mainstream technology it is today.

A recent book, *Patterns for Parallel Programming* [69], is an attempt to do for parallel programming what the GOF book did for object-oriented programming. Unlike the GOF book, the *Parallel Patterns* book is more than a catalog of patterns. It is a pattern language — an interconnected web of patterns designed to capture a way to think about parallel programming. The authors organized their pattern language around four layers:

- Find the concurrency in your problem.

- Develop a strategy to exploit the concurrency.

- Find and use the appropriate algorithm pattern.

- Find the supporting patterns to help you implement the algorithm in software.

We will use these same layers in our own discussion producing a framework of patterns to describe the process of constructing parallel algorithms. We will use many of these patterns in the rest of the book as we consider how different programming languages support the different algorithm patterns.

9.2 Finding concurrency

As we discussed earlier, the foundation of any parallel program is concurrency. There must be sequences of instructions that can effectively run at the same time in order to have a parallel program. The challenge to parallel programmers is to identify the concurrency in the problem they wish to solve, restructure the problem to expose the concurrency, and finally to exploit the concurrency with a parallel algorithm.

Finding the concurrency in a problem can be more art than science. The key is to look at a problem and think of different ways to decompose the problem to create useful concurrency. We must then consider the different potential decompositions and explore how each maps onto different parallel algorithms. Finally, we can evaluate the decompositions under different metrics (such as performance or programmability), and pick one that seems to work best for the problem at hand.

The process starts with the basic decomposition of the problem using these steps:

1. Identify sequences of instructions that can execute as a unit and at the same time. We call these *tasks*.

2. Decompose data operated on by the tasks to minimize overhead of sharing data between tasks.

3. Describe dependencies between tasks: both ordering constraints and data that is shared between tasks.

For example, if you have a thousand files to search to find the frequency in which different patterns occur, the search operations on each file define the tasks. The data is decomposed by assigning each file to a task. Data dependencies arise from the shared data structure that keeps track of which patterns are found and how often they occur.

The fundamental insight is that all parallel algorithms require you to find tasks that can execute at the same time, and that you decompose the data so the tasks can run as independently as possible. There is always a task decomposition and a data decomposition.

Where the art comes into the process is which decomposition should drive the analysis. For example, there are many problems, typically called data-parallel problems, where the data decomposition is so straightforward that it makes sense to define the tasks in terms of the data. For example, if we are applying a filter that replaces each point in an image with an average computed from a set of neighboring points, the parallelism is easy to define in terms of a tiled decomposition of the image. The data decomposition is the set of tiles from the image, the tasks are the update of each tile, and the data dependencies are the boundaries of the tiles needed to update the average values at the edges.

Analyzing the program to split up work into tasks and associate data with these tasks requires us to revisit the fundamental topics discussed throughout the earlier portion of this text. For example, dependencies play a critical role in identifying what computations must be performed before others. These dependencies can be both logical dependencies that dictate the order that specific operations must be performed in, or data dependencies dictating an order in which data elements must be updated. In most programs, we will not be able to break the problem into a set of purely independent tasks and data elements. This leads to the need for coordination and synchronization. Recognizing dependencies and identifying the coordination and synchronization mechanisms necessary to decompose a problem into a correct concurrent implementation is the core task of a parallel programmer.

9.3   Strategies for exploiting concurrency

Once a problem is understood in terms of its tasks, data decomposition and dependencies, you need to choose a strategy for exploiting the concurrency. A common approach (as discussed in [69] and [17]) is to develop a strategy along one of three options:

1. The collection of tasks that are to be computed: *agenda parallelism.*

2. Updates to the data: *result parallelism.*

3. The flow of data between a fixed set of tasks: *specialist parallelism.*

Problems that map well onto agenda parallelism generally have a well defined set of tasks to compute. They may be a static set (such as a fixed set of files) or generated through a recursive data structure (defining tasks as subgraphs exposed through a graph partitioning process). The common theme is that the programming language constructs are used in the source code to define the tasks, launch them and then terminate the algorithm once the tasks complete.

In result parallelism, the algorithm is designed around what you will be computing, i.e., the data decomposition guides the design of the algorithm. These problems revolve around a central data structure that hopefully presents a natural decomposition strategy. For these algorithms, the resulting programs focus on breaking up the central data structure and launching threads or processes to update them concurrently.

Finally, specialist parallelism occurs when the problem can be defined in terms of data flowing through a fixed set of tasks. This strategy works best when there are a modest number of compute intensive tasks. When there are large numbers of fine grained tasks, its usually better to think of the problem in terms of the agenda parallelism strategy. An extremely common example of specialist parallelism is the linear pipeline of tasks, though algorithms with feedback loops and more complex branching structure fit this strategy as well.

Once a strategy has been selected, its time to consider the specific algorithms to support concurrency within the selected strategy.

9.4 Algorithm patterns

With decades of parallel programming experience to work with, a rich variety of parallel patterns have been mined, which is the core topic of [69]. Based on the strategy chosen to exploit concurrency for your parallel program, you can quickly narrow in on the appropriate algorithm pattern to use.

Agenda parallelism

For agenda parallelism, the pattern naturally focuses on the tasks that are exposed by the problem. The following patterns, most documented fully in [69], are commonly used for the agenda parallelism strategy:

- *Task parallel*: The set of tasks are defined statically or through iterative control structures. The crux of this pattern is to schedule the tasks so the balance is evenly spread between the threads or processes and to manage the data dependencies between tasks.

- *Embarrassingly parallel*: This is a very important instance of the task parallel pattern in which there are no dependencies between the tasks. The challenge with this pattern is to distribute the tasks so the load is evenly balanced among the processing elements of the parallel system.

- *Separable dependencies*: A subpattern of the task parallel pattern in which the dependencies between tasks are managed by replicating key data structures on each thread or process and then accumulating results into these local structures. The tasks then execute according to the embarrassingly parallel pattern and the local replicated data structures are combined into the final global result.

- *Recursive algorithms*:   Tasks are generated by recursively splitting a problem into smaller subproblems. These subproblems are themselves split until at some point the generated subproblems are small enough to solve directly. In a divide and conquer algorithm, the splitting is reversed to combine the solutions from the subproblems into a single solution for the original problem.

Result parallelism

Result parallelism focuses on the data produced in the computation. The core idea is to define the algorithm in terms of the data structures within the problem and how they are decomposed.

- *Data parallelism*:   A broadly applicable pattern in which the parallelism is expressed as streams of instructions applied concurrently to the elements of a data structure (e.g., arrays).

- *Geometric decomposition*:   A data parallel pattern where the data structures at the center of the problem are broken into subregions or tiles that are distributed about the threads or processes involved in the computation. The algorithm consists of updates to local or interior points, exchange of boundary regions, and update of the edges.

- *Recursive data*:   A data parallel pattern used with recursively defined data structures. Extra work (relative to the serial version of the algorithm) is expended to traverse the data structure and define the concurrent tasks, but this is compensated for by the potential for parallel speedup.

Specialist parallelism

The specialist parallelism strategy can be pictured as a set of tasks that data flows through. Large grained data flow methods are a good example of this pattern in action.

- *Pipeline*:   A collection of tasks or pipeline stages are defined and connected in terms of a fixed communication pattern. Data flows between stages as computations complete; the parallelism comes from the stages executing at the same time once the pipeline is full. The pipeline stages can include branches or feedback loops.

- *Event-based coordination*:   This pattern defines a set of tasks that can run concurrently. Each task waits on a distinct event queue and executes as events are posted.

Together the above patterns represent the most commonly used approaches to parallel application design. This is not a complete list; completeness being all but impossible. Programmers constantly invent new patterns or mix them together in new ways. This is the strength of patterns; they capture the essence of a solution to a common class of problems, but they are flexible enough to permit creative expression into an endless variety of algorithm designs.

9.5 Patterns supporting parallel source code

We can also define patterns that sit below the algorithm patterns and play a supporting role defining how the algorithms are expressed in source code. In [69] these are the *supporting structure* and *implementation mechanism* patterns. Unlike the parallel algorithm patterns, these patterns are fundamentally about how an algorithm is expressed in source code. Consequently, it can be useful to keep these patterns in mind as we study how different programming languages interact with the algorithm patterns.

Fork-join

A computation begins as a single thread of control. Where needed, concurrency is added by creating a team of threads (forking) that execute functions which define the concurrent tasks. When the tasks complete and terminate (join), the computation continues as a single thread. A single program may contain multiple fork/join regions. In some cases, they may even be nested.

SPMD

Multiple copies of a single program are launched. They are assigned a unique contiguous ID (a rank). Using the ID and the total number of programs as parameters, the pathway through the program or the associated data is selected. This is by far the most commonly used pattern with message passing APIs such as MPI.

Loop parallelism

In this pattern, parallelism is expressed in terms of loops that are shared between a set of threads. In other words, the programmer structures the loops so iterations can execute concurrently and then directs the compiler to generate code to run the loop iterations in parallel.

Master-worker

This pattern is particularly useful when you need to process a collection of tasks but the time needed for each task is difficult to predict. A process or thread is assigned the role of *master* while the other threads or processes become *workers*. The master sets up a task queue and manages the workers. The workers grab a task from the master, carry out the computation, and then return to the master for their next task. This continues until the master detects that a termination condition has been met, at which point the master ends the computation. This pattern is conducive to easily implementing methods to automatically balance the load between a collection of processing elements.

SIMD

This data parallel pattern rigorously defines the computation in terms of a single stream of instructions applied to the elements of arrays of data. Vector instructions found in modern microprocessors are a common example of a SIMD API. SIMD is one classification of parallel execution put forth by Flynn (see section 6.1.8).

Functional parallelism

Concurrency is expressed as a distinct set of functions that execute concurrently. This pattern may be used with an imperative semantics in which case the way the functions execute are defined in the source code (e.g., event based coordination). Alternatively, this pattern can be used with declarative semantics, such as within a functional language, where the functions are defined but how (or when) they execute is dictated by the interaction of the data with the language model.

9.6 Demonstrating parallel algorithm patterns

Now that we have a framework to describe the process of parallel algorithm design and have identified some important design patterns, we can consider how the patterns interact with different high-level parallel programming languages. This will be the topic of the next four chapters. The patterns we will investigate include:

- Task Parallelism
- Data Parallelism
- Recursive Parallelism
- Pipelining

To make the discussion more concrete, we will use a few modern languages that contain concurrency features in our discussion. In the appendices we provide a brief and high-level overview of the languages that we draw from in our discussions.

9.7 Exercises

1. Pick a common design pattern from sequential programming. Describe the abstraction that the pattern embodies and how you would apply the pattern when implementing an algorithm. Refer to [37] or the Web for references to common sequential patterns if you are unfamiliar with them.

2. Consider a real-world activity that can take advantage of concurrency, such as cooking a meal or assembling an automobile. In section 9.2, we give three steps to decompose a problem and understand where potential concurrency exists. Apply this to the real-world activity you have chosen and describe what you find at each step.

3. Consider a computation to render a 3D model to update a frame to be displayed on computer's monitor. Identify the concurrency in this problem (section 9.2). Describe how this concurrency could be exploited using each of the strategies defined in section 9.3.

4. Computer games and other consumer applications utilize multiple patterns in a single program, often in hierarchical ways. Pick an application program you know well and describe how it might use the algorithm patterns from section 9.4.

5. Pick a target computer system such as a CPU/GPU heterogeneous platform or a multi-core processor. Pick a set of supporting patterns from section 9.5 to support the algorithm patterns used for the application you selected for Problem 3. How would these patterns change if you considered a radically different architecture such as a cluster?

Chapter 10

Pattern: Task Parallelism

Objectives:

- Introduce the task parallel pattern and supporting algorithmic structures.
- Discuss two examples of the task parallel pattern applied for implementing genetic algorithms and computing the Mandelbrot set.
- Demonstrate these examples using Erlang, Cilk, and OpenMP.

We will start our discussion of patterns for parallel programming with that which is likely most familiar — *task parallelism*. The task parallelism pattern is one of the most common design patterns in parallel computing with applications including ray tracing, sampling large parameter spaces in simulations, or processing long collections of distinct data.

Task parallelism is very close to the way we achieve concurrency in the real world. If we have, for example, ten thousand envelopes to stuff and address, we can recruit a team of 100 workers, giving each 100 envelopes to manage. If the workers are all equally efficient and we keep overheads low, we can hope to speed up the process 100 fold. This is a classic example of the task parallel pattern.

Let's consider the task parallelism pattern in more detail. A task is a logically connected sequence of statements in a program. Our challenge is to define the tasks and manage them so they execute to effectively implement the overall program both correctly and efficiently. The essential elements of the solution to a task parallelism problem are to:

1. Define the set of tasks that make up our algorithm.

2. Manage dependencies between the tasks.

3. Schedule their execution so the computational load is evenly balanced among processing elements.

4. Detect the termination condition and cleanly shut down the computation.

When the tasks are completely independent, the problem is called *embarrassingly parallel*. In this case, the programmer just defines the tasks and schedules them for execution. Techniques are well known for creating algorithms that automatically balance the load between processing elements for embarrassingly parallel problems. We will see this used in our parallel implementation of the Mandelbrot set computation to keep the load balanced and avoid having idle processors.

When dependencies are present, the application of this pattern is more complicated. Often, the dependencies are *separable* and can be "pulled out" of the tasks converting the problem into an embarrassingly parallel problem. A common method for dealing with separable dependencies is to use *data replication* followed by a *global recombination* process. For example, if the tasks carry out a computation and accumulate statistics on the results, the data structures used to hold accumulated results are replicated; once for each processing element in the computation. Once all the tasks are complete, the local copies of the data structures are combined to produce the final result. In essence the problem is converted into an embarrassingly parallel pattern.

More complex applications of the task parallel pattern emerge when the dependencies are not separable. In such cases, the concurrent updates must be explicitly managed using a shared data structure pattern (distributed array or shared queue) or through explicit mutual exclusion protocols. As the dependencies become more complex, the pattern becomes more difficult to apply and at some point a data parallel pattern or geometric decomposition becomes a more natural pattern to work with.

10.1   Supporting algorithm structures

There are many ways to support the task parallel pattern in software. Commonly used design patterns from [69] include master-worker, SPMD, loop-level parallelism and fork-join control flow.

Master-worker

A master manages a collection of tasks (often with a queue) and schedules them for execution on by a collection of workers. Workers request more tasks as they finish their assigned work, the result being that the load is automatically balanced between a collection of workers by those that complete quickly taking additional work instead of sitting idle.

SPMD

A single program is run by multiple processing elements. They use the ID of each processing element to select between a set of tasks and to manage any shared data structures. This design pattern is very general and not specific to task level parallelism. MPI programmers use this pattern.

Loop-level parallelism

Collections of tasks are defined as iterations of one or more loops. The loops are divided between a collection of processing elements to compute tasks concurrently. This design pattern is also heavily used with data parallel design patterns. OpenMP programmers commonly use this pattern.

Fork-join

Tasks are associated with threads. The threads are spawned (forked), carry out their execution, and then terminate (join). Note that due to the potential high cost of thread creation and destruction, programming languages that support this pattern often use logical threads that execute with physical threads pulled from a thread pool. But to the programmer, this is an implementation detail that the compiler and runtime author is concerned with. Cilk and the explicit tasks in OpenMP 3.0 commonly use this pattern.

10.1.1 The Master-worker pattern

Of the set of supporting patterns, the master-worker pattern is most closely associated with the task parallel pattern. Even when the SPMD or loop-level patterns are used, the basic idea of the master-worker pattern is often exploited but with the master implied by the behavior of the scheduling algorithm. The master-worker pattern [69] is extremely powerful. When applied carefully, it has the feature of automatically balancing the load between a collection of workers. This makes it one of the most straightforward ways to support dynamic load balancing.

The essential elements of this pattern are:

- *Master*: A single thread that manages a set of tasks and coordinates the execution of the workers. In a common variation of this pattern, the master sets up the computation and then becomes a worker. This allows cycles to be used to perform useful work that would otherwise be wasted while the master waits for workers to complete.

- *Workers*: One or more threads that compute the tasks. They grab a task from the master, carry out the computation, and then return to the master to pick up the next task. It is this dynamic scheduling capability that results in the dynamic load balancing. It should be noted that static scheduling can be used in some master-worker programs, in which the tasks assigned to each worker are predetermined and not dynamic. Load balancing is harder to achieve when static scheduling is used.

The tasks must be independent, or if there are dependencies, they must be separable dependencies that can be locally managed by the workers and collected at the end of the computation to produce the global result. The collection phase that combines local results into a global result can be implemented efficiently. Reductions in OpenMP and inlets in Cilk are examples of language-level mechanisms to perform this combination step.

In pseudo-code, we can think of the master and worker having the structure shown in Listing 10.1.

In this simplified case, the master is responsible for generating task objects, adding them to a queue of pending tasks, and then accumulating the results as they return. Notice that there is no explicit relationship between the order in which tasks are generated relative to the order that results are returned. This is common in this pattern, and it is often the responsibility of the master to perform any additional work necessary to stitch the individual results into the global result that the set of tasks was seeking to accomplish (should such a global result exist). The worker is quite simple. It simply waits for tasks to arrive in the queue, invokes the logic necessary to generate a result from the task specification, and pushes this result into the result set visible by the master.

A correct use of the master-worker pattern must address the following issues:

- Create a master and one or more workers as concurrent threads or processes.

- Create a set of tasks and concurrent data structures to manage them. In particular, all tasks must execute *at most once* and without race conditions.

- Collect results from the workers without introducing data races.

- The master needs a way to detect a terminal condition and cleanly shut down the workers.

10.1.2   Implementation mechanisms

There are many ways to implement task parallel programs. A distinguishing feature between task-parallel implementations is the persistence of the concurrent workers. Some systems require that workers be created for each task as the tasks are created, and disposed of when the tasks are complete (e.g., Cilk with the fork-join pattern). Other systems assume a persistent, fixed set of workers based on the processing resources that are available and

Listing 10.1: Pseudo-code for a master-worker program.

```
//-------------------------------------------------
// master task
master {
  SharedQueue tasks;
  Results R;
  int i, taskCount;

  // First, generate the set of tasks and
  // place them in the shared queue.
  //
  for (i=0;i<taskCount;i++) {
    Task t = generateSubTask();
    tasks.add(t);
  }

  // Next, consume the set of results produced
  // by the workers.
  //
  for (i=0;i<taskCount;i++) {
    Result res = R.pop();
    consumeResult(res);
  }
}

//-------------------------------------------------
// worker task
worker {
  SharedQueue tasks;
  Results R;

  while (true) {
    // Get a task
    task = tasks.pop();

    // Perform the work
    Result res = performWork(task);

    // Put the result in the result queue
    R.push(res);
  }
}
```

provide a queueing and scheduling system to handle giving out work to and organizing results from the set of workers (e.g., the master-worker design pattern). Erlang is an example where workers can be implemented as persistent processes. In either case, the logical decomposition of the program into tasks is an essential defining feature of the implementation.

The choice of a fixed set of persistent workers versus transient workers that come and go is often application or language dependent. In languages that provide a thread abstraction above the underlying runtime implementation, a programmer may program as though threads are transient while the language and runtime draw from a pool of persistent threads to do the actual work. OpenMP and Java (via thread pool objects) are two common examples that use this model. On the other hand, if the programmer is working directly with the underlying threads, then care must be taken to avoid creating more threads than necessary to avoid contention between them and subsequent performance degradation (e.g., direct use of a low-level thread library such as POSIX threads).

In addition to defining workers and sets of threads, the programmer needs to manage the interaction between the units of execution working on the tasks. This is often left to the programmer who must create and manage concurrent data structures. Some early languages that made heavy use of the master-worker pattern provided a distinct memory region to help coordinate the workers and sets of tasks. The notion of "tuple-spaces" in Linda [17] and more recently the JavaSpaces [36] package from Sun Microsystems are examples.

10.1.3   Abstractions supporting task parallelism

To implement a task parallel program, we must be able to define the tasks to perform, the subprograms that perform them, and the coordination and communication mechanism necessary to distribute tasks and gather their results to form the overall result of the program.

Worker creation

Languages that support threads typically contain keywords or intrinsic functions to spawn new threads and identify the threads for purposes of communication and control. For example, we can spawn processes in Erlang using the `spawn` operation.

```
Pid = spawn(fun loop/0).
```

This will spawn a new process that can identified by `Pid` which executed the function `loop` as its body. Tasks can be sent to the new process by passing messages to the `Pid`, and results can later be sent as messages to the master process. Similarly, the `spawn` keyword in Cilk is available.

```
spawn func();
```

This will create a thread which executes the Cilk function `func` as its body. Unlike Erlang, Cilk does not generate an identifier to allow the parent to identify and interact with the child. To achieve this in Cilk, we would have to pass an identifier as an argument to the thread that would be used to identify some data element in the shared memory between the two that can be used for communication. For many problems where the task representations require only a moderate amount of storage space, the communication of the task to the worker is achieved through arguments passed to the spawned function, and results are returned to the master as the results of the function. The master can use inlets to perform any necessary synchronized logic to stitch the results from each worker into the final global result.

In Java, we can achieve this using the thread mechanisms provided as part of the standard library. A class can be constructed that implements the `Runnable` interface (or extends the `Thread` class) to represent the thread. The functionality of the thread is then placed in the `run()` method of the object. The thread then can be started manually (via the `start()` method), or via one of a set of `Executor` objects that are provided by the Java standard library. These executors are used to transparently implement thread pools and other novel thread scheduling mechanisms, and can be customized should the programmer require it. The significant difference between what we see in Java versus languages like Cilk or Erlang is that the Java functionality is provided by a library. The classes and interfaces that are used to implement threads are part of the standard library, and are not as tightly integrated into the syntax of the language as the spawning mechanisms in parallel languages.

Worker interaction

Given a set of worker threads, a task parallel program is based on tasks being handed to these workers and their results being gathered and integrated into the final end product. One simple method to implement this is to pass the data necessary for the task (or some information used to find it) in when the worker is created. In Cilk this is easy to achieve by passing the task information as arguments when the worker is spawned. We can also pass in a pointer to the location where the worker should place the results of its computations so the parent thread can find it.

```
spawn worker(task_data, &results);
```

Due to the asynchronous nature of Cilk threads, the parent must synchronize with the workers before attempting to use their results. For example, say we have a function that we wish to execute in five threads, each of which will take its own task data as input and place a single integer in a result slot that is passed to it by reference.

```
int results[5], data[5], i;

/* spawn set of workers to execute
   concurrently */
for (i=0;i<5;i++)
    spawn worker(data[i],&results[i]);

sync; /* wait for threads */

/* safe to work with results here */
```

Without the sync, it is possible that the parent will attempt to use a value that has not yet been placed in the results array, likely resulting in incorrect computations being performed by the master thread.

Erlang uses a message passing abstraction to allow threads to interact without the use of a shared memory. This helps prevent problems that we can encounter with shared memory programs where synchronization isn't properly implemented. Given an identifier for a process returned by the spawn operation, another process can send a message to the thread using the send operator. If Pid is the process identifier, and TaskData is a variable representing the data that the corresponding task is to operate on, sending the data to the process is quite straightforward.

```
Pid ! TaskData.
```

On the other side of the messaging interaction, the receive operation is used to take messages from other processes and invoke operations on them within the receiving process.

```
receive
    SomeData -> func(SomeData)
end.
```

If we want to have an interaction in which the worker is able to later respond to the master process, then we must explicitly send along with the data the process identifier of the master. This is quite simple and achieved by adding the process identifier of the sender, determined via the self() call, along with the data sent to the worker.

```
Pid ! {self(), TaskData}.
```

The worker can then use this to send the result back to the master when it finishes.

```
receive
    {MPid,SomeData} -> MPid ! {self(), func(SomeData)}
end.
```

How do we read this new line of code? First, any message that is received of the form {MPid, Data} from the process MPid results in the operation on the right hand side of the arrow being executed. The right hand side is actually two steps rolled into one. First, we create a new message to send back to the master process composed of the worker's process ID (self()), since the worker needs to send its own ID) and the result of evaluating the function func() on the data that was received (SomeData). When func completes and the message is fully formed, it is then sent to the master using the identifier MPid that arrived with the original task request. Note that we haven't written the necessary code to deal with messages that do not match this form. Programmers will typically provide such code to gracefully deal with messages that do not have the expected structure.

We will now discuss two case studies that are easily mapped onto the task parallel algorithm pattern, using Erlang and Cilk as our example languages.

10.2 Case study: Genetic algorithms

Genetic algorithms are a form of evolutionary algorithm used to solve what are known as optimization problems. An optimization problem is one in which a function known as an *objective* is minimized or maximized. This means that the algorithm will identify some *optimal solution* that when evaluated with the objective function will yield a value that is "more optimal" than other candidate solutions. In the case of minimization, an optimal value will evaluate to the lowest value of all possible candidates for the objective function. One of the simplest examples of this that many people encounter is what is known as linear regression. Say we have a set of points $P = (x_i, y_i)$, and we wish to find the parameters a and b of the function $f(x) = ax + b$ such that

$$\sum_{i=1}^{|P|} |f(x_i) - y_i|$$

is minimized. A perfect fit would yield a value of zero for the summation.

One approach to solving this problem is to pick arbitrary values of a and b, and evaluate the objective function with those parameters to see how well it does. We can then modify them slightly and evaluate nearby values of a and b to see if any can do better, and iterate this process until we reach values that we are satisfied with. For example, if we try the values 1 and 2 for a, and see that for some x, $f(x)$ is closer to the value we are seeking for $a = 2$ than $a = 1$, we can infer that the optimal value for a is likely to be closer to 2 than 1, and maybe we will choose 2 and 2.5 on the next iteration. We illustrate this

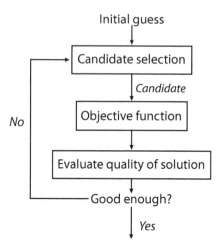

FIGURE 10.1:   A simplified look at the components and steps making up a basic iterative optimization algorithm.

iterative process of selecting parameters and evaluating the objective function in Figure 10.1.

Of course, in simple cases like this, there are well defined mathematical methods to determine the regression parameters without a search through parameter space. Genetic algorithms and other sophisticated optimization techniques exist to deal with complex objective functions that can take potentially huge numbers of parameters and have no clear analytical method to determine their optimal values. Furthermore, complex problems can make the selection of new candidates as the algorithm iterates more complex than the numerical regression example above. Genetic algorithms are one instance of optimization technique that can tackle these complex problems.

From an algorithmic perspective, many optimization techniques are good candidates for implementing as task parallel programs. The process of selecting candidates and evaluating them does not need to be performed one at a time. In fact, it is often desirable to evaluate a set of candidates at the same time to understand the properties of the objective function with respect to variations in parameters. We can think of the set of candidates as forming a type of landscape, where the quality of the solution corresponds to the height of the data point. The optimization algorithm is attempting to find a peak or valley in this landscape with no map, so it is trying to use a disciplined yet blind method to search. Clearly one can build a picture of this landscape and potentially reach the desired optimal value faster if one is able to probe multiple candidates at the same time, effectively looking out in many directions all at once.

The algorithm we will implement here is based on the *gene expression pro-*

gramming evolutionary algorithm [34]. The components of the algorithm are:

- The objective function, which takes a candidate solution as input and produces a numerical measure of the quality, or *fitness*, of the solution applied to the objective function.

- A candidate selection algorithm that takes knowledge of those tried so far and attempts to build better candidates to evaluate.

- A method to turn encoded individuals represented by a chromosome of characters into a form that can be evaluated, such as an arithmetic expression.

- A set of operators that are used to generate new populations from old ones based on the fitnesses of the old individuals.

To parallelize this, we can replicate the objective function (which can be quite computationally intensive for some problems) across multiple parallel processing elements, and set up the candidate selection algorithm to produce multiple candidates to evaluate in each pass through the optimization loop. Thus we have a master process that is responsible for generating individuals to be evaluated and synthesizing its fitnesses to build new generations to evaluate, and worker processes that are responsible for evaluating the fitness of individuals.

In genetic algorithms, candidate solutions to the problem are known as individuals, and the set that the algorithm evaluates during each generation is the population. The act of selecting new candidates for evaluation is the phase of reproduction, and is based on the application of genetic operators to the individuals selected from the population based on its fitness to build a new population. Individuals that performed well on the objective function are more likely to persist than those that performed poorly (the very worst of which will simply be eliminated), and new individuals will be created by combining parts of successful ones in hopes that a better individual will result.

For this task, we will write our program in Erlang. Writing this in Erlang is actually quite easy. Most of the effort is actually related to writing the sequential operations that make up the genetic algorithm, such as the genetic operators to create new individuals and the methods to express the chromosomes of each individual into a form that can be evaluated. The code for these operations is provided on the Web site for this book, so as not to distract from the discussion of the concurrent aspects of the algorithm here. We chose Erlang for this example since a message-passing abstraction fits well with the master-worker concurrent implementation, and Erlang has a particularly concise syntax for this purpose.

We start by structuring the program in two major components.

Master: The master process is responsible for managing the population, which entails collecting fitness information for each individual and using this to generate new populations by applying genetic operators during selection and reproduction.

Worker: The worker processes receive a subset of individuals from the master, and are responsible for converting them from their coded chromosome form into a form that can be executed. This is known as *expression* of the chromosomes. The worker will then evaluate each individual against the set of test values to determine its fitness, and when it is complete, it will send the fitness values back to the master.

10.2.1   Population management

Four functions make up the core functionality of the master process with respect to delegation of subpopulations to workers to evaluate and the accumulation of the evaluated fitnesses as they complete. We show the four functions in Listing 10.2.

make_pids

The first function of interest is `make_pids`. This function takes as input the number of workers to create, spawns them, and returns a list of their identifiers. The function that each worker implements, `worker:loop` is described below in section 10.2.2. The main routine of the driver (not shown here) will invoke `make_pids` to create the set of workers, and then pass that to the `parallel_fitness_eval` function which hands subsets of the population off to each worker.

parallel_fitness_eval and sublists

The `parallel_fitness_eval` function performs three important operations. Before it gets to these, it first counts the number of workers and individuals for convenience purposes. The first operation of interest that it performs is a call to the `sublists` function. This function takes the number of worker processes, the number of individuals allocated to each worker, and the full set of individuals as input. As output, it returns a list of lists, each of which represents the subpopulation corresponding to a single worker. This represents the creation of tasks that are to be executed concurrently.

The second major operation this function performs is the communication of the subpopulations to the worker threads. Up to this point, we have a list of worker identifiers, and a list of subpopulations, both of which have the same length. In Erlang, we cannot address lists like we would arrays in other languages — it isn't efficient to ask for the ith member of a list, as that would be an $O(i)$ complexity operation instead of the desired constant time that arrays would provide. So, we use an operation known as a *zipper*, which takes two lists of N elements, and creates a new list containing N elements where the ith element is a pair representing the ith element from each of the

Listing 10.2: Master functions.

```
%
% spawn a set of worker processes
%
make_pids(0) -> [];
make_pids(N) -> [(spawn(fun worker:loop/0))]++
                (make_pids(N-1)).


%
% make sublists for workers
%
sublists(1,     _, L) -> [L];
sublists(N, Size, L) ->
  {A,B} = lists:split(Size,L),
  [A]++(sublists(N-1,Size,B)).


%
% delegate subpopulations to workers
%
parallel_fitness_eval(Pids,Population,MaxFit,
                      DataPoints,DataValues) ->
  NumPids = length(Pids),
  NumIndividuals = length(Population),
  SubPops = sublists(NumPids,
                     round(NumIndividuals/NumPids),
                     Population),
  lists:map(fun({I,Pid}) ->
            Pid ! {self(),
                   {work,
                    {I,MaxFit,DataPoints,DataValues}}}
            end,
          lists:zip(SubPops,Pids)),
  gather_results(NumPids,[],[]).

%
% gather evaulation results from workers
%
gather_results(0,R,P) -> {R,P};
gather_results(N,R,P) ->
  receive
    {Pid,{Results,SubPop}} ->
      gather_results(N-1,R++Results,P++SubPop)
  end.
```

original lists. This zip operation allows us to generate pairs where the worker identifier and the corresponding subpopulation are together in a single data element.

To send the data, we iteratively apply (or *map*) the function to each of these pairs. Given the identifier of a worker and the subpopulation, we use the send operator (!) to pass the subpopulation to the waiting process. The message that we send encodes the identifier of the master using the self() call. In addition to the subpopulation, we also pass the set of data points where each individual is to be evaluated to quantify its fitness and the maximum fitness value (MaxFit) that an optimal individual will achieve.

gather_results

Finally, the routine gather_results is invoked. This function is passed the number of workers, and two empty lists that will eventually be grown to contain the fitness results from the concurrently executing workers. The gather_results routine blocks to receive a message from a single worker, where the message contains the set of fitness values and the corresponding set of individuals. When a message is received, the newly acquired results and individuals are concatenated with the set of results and individuals already received, and the function is recursively called with a decremented worker count. When the call is made with a worker count of zero, the routine returns the accumulated results and individuals.

The point of accumulating up both the results and the individuals is due to the unknown order in which the subpopulation evaluations will occur. It is quite possible that the workers will complete in an order different than that in which their work was initially distributed. As such, the master must record the fitness results and the individuals they correspond to in the correct order. Once the gather_results function completes, the parallel_fitness_eval function will return and the master can do its local sequential task of selection and reproduction so that this process can be repeated until a sufficiently optimal solution is found.

10.2.2 Individual expression and fitness evaluation

The task executed in a worker is much simpler than that within the master. The code that interacts with the master for the worker is shown in Listing 10.3. The core of the worker is the loop function, which executes a blocking receive waiting for incoming messages. When a message arrives containing the sender (From) and the task body, the worker evaluates each individual in the subpopulation that it received using the sequential eval_fitness_individual function. This function converts the encoded individual into one that can be executed, and evaluates it with the provided data points and expected values to determine its fitness. This iterative evaluation of the individuals is achieved with the map primitive. When all individuals have been evaluated, the worker

Listing 10.3: Worker main function.

```
loop() ->
  receive
    {From, {work, {Pop,M,DataPoints,DataValues}}} ->
      Results = lists:map(fun(I) ->
          driver:eval_fitness_individual(I,M,
            DataPoints,DataValues) end, Pop),
        From ! {self(), {Results,Pop}},
        loop();
      {From, Other} ->
        From ! {self(), {error,Other}},
        loop()
    end.
```

sends a message back to the master that contains the identifier of the worker process, and a pair containing the list of results and the corresponding individuals that they were derived from.

This loop routine also checks for messages that do not match the expected format. In this case, the result returned to the sender would be an error with the erroneous message payload (**Other**) included. In both cases within the receive, the function recursively calls itself to repeat the process and wait for the next message to work on. The code as written must be terminated externally (such as killing the Erlang process).

10.2.3 Discussion

This section has demonstrated how we can take a problem and decompose it into a set of tasks that can be delegated to a set of workers by a master process. We used message passing as the means for master-worker interactions. In this example, we used a static decomposition of tasks to workers — each worker received a fixed subpopulation. If the individuals in one subpopulation evaluated faster than those in another, we would see some workers spending time idle. In the next example, we will see how dynamic scheduling of the pool of tasks to a set of workers can provide load balance in problems where an imbalance of work per task is expected.

FIGURE 10.2:   A picture of the Mandelbrot set.

10.3   Case study: Mandelbrot set computation

One of the classic programs that students build for learning parallel computing has been the computation of the familiar Mandelbrot set fractal shown in Figure 10.2. We will illustrate in this section how the problem can be decomposed into a set of tasks that are implemented using a master-worker pattern. This problem is also a convenient case to study the load balancing properties that come easily with the master-worker model. In this example, we will use the parallel language Cilk as our language of choice.

10.3.1   The problem

The problem that we will solve is that of generating an image of the Mandelbrot set. The Mandelbrot set is defined as a set of points in the complex plane where each point has associated with it a value representing how long an iterated function on that point takes to exceed a certain value. More precisely, given some complex number c, we define a relation

$$z_{n+1} = z_n^2 + c. \tag{10.1}$$

The starting value of the iteration, z_0, is set to the complex number zero. For each point c in the complex plane where we wish to compute Eq. 10.1, we iterate this equation starting with c and z_0, computing z_1, z_2, and so on. The problem becomes interesting when we pick two constants — a threshold "escape" value (call it t) and a maximum iteration count (n_{max}). For each point c that we are interested in, we repeat the iteration of Equation 10.1 until

Listing 10.4: Main loop of the Mandelbrot set computation for the whole image.

```
for (i=0; i<N; i++) {
  for (j=0; j<N; j++) {
    I[i][j] = count_iterations(a + i*dx, b + j*dy);
  }
}
```

either $|z_{n+1}| > t$ or $n = n_{max}$. In both cases, we record the value of n for the point c. When this is repeated for a set of points in a small region of the complex plane, we see the well known visualization of the set by associating a color with the iteration counts recorded for each point.

How do we implement this? Before we tackle the parallel implementation, we must understand the sequential implementation. We start by picking a rectangular region of the complex plane over which we wish to compute the set. Remember that a complex number z is defined as $z = x + yi$ where x is known as the real component, y as the imaginary component, and i as the imaginary unit value $\sqrt{-1}$. As such, complex numbers are treated as vectors representing coordinates in the two-dimensional complex plane with one axis defined by the real component and the other by the imaginary component. Thus we can pick two complex numbers representing two opposite corners of our rectangle. Call these $U = a + bi$ representing the upper left corner, and $L = c + di$ representing the lower right.

Now, to create an image, we want to discretize this rectangle into a set of pixels. To simplify the problem, let us assume the image will be square with N pixels to a side. Thus we are partitioning the plane defined by U and L into points that are separated by $dx = \frac{a-c}{N}$ in the horizontal direction and $dy = \frac{b-d}{N}$ in the vertical. Given this information, we can write the loop that iterates over the image pixels storing the iteration counts as defined above in the 2-D N-by-N array I as shown in Listing 10.4. The function count_iterations that iterates Equation 10.1 is defined in Listing 10.5.

If we wrap the nested loops that iterate over I in a suitable main routine that allocates I and sets up the values dx, dy, nmax and t, our sequential program would be complete. Now we can think about decomposing it into a task parallel program.

10.3.2 Identifying tasks and separating master from worker

As repeatedly discussed in the book so far, we know that identifying dependencies within a program is critical to identifying parallelism. In this example, we are fortunate to see that each iteration of the loop that populates I is independent of the others, so there is a high degree of parallelism available

Listing 10.5: Function to iterate Eq. 10.1 for a single point.

```
int count_iterations(double r, double i) {
    int n;
    double zr = 0.0, zi = 0.0;
    double tzr, tzi;

    n = 0;
    while (n < max_iters && sqrt(zr*zr + zi*zi) < t) {
        tzr = (zr*zr - zi*zi) + a;
        tzi = (zr*zi + zr*zi) + b;
        zr = tzr;
        zi = tzi;
        n = n+1;
    }

    return n;
}
```

within that loop. On the other hand, the loop within the count_iterations routine has a dependency in which each iteration requires the results of the previous one. This limits the degree to which we can exploit parallelism in that code.

We will focus our attention on the loops that populate I where the independence of each iteration makes parallelization easy. The highest amount of parallelism we could achieve would be to execute the computation for each pixel in parallel. This, unfortunately, is not likely to be feasible as we would require N^2 processing elements to actually implement this. On the other hand, if we have P processing elements available, we can divide the image into blocks of size $\frac{N^2}{P}$, and let each block be computed concurrently. Unfortunately, this naive approach to parallelizing the problem is also not ideal, although for a subtle reason.

Consider for a moment what the Mandelbrot set is a visualization of. Each pixel is assigned a color based on the number of iterations that were required for the value it corresponds to in the complex plane to either exceed t or the maximum iteration count. The fact that we see intricate patterns in the colors when we look at the set means that there is a high degree of variation in the number of iterations required for each pixel. Therefore we can conclude that the time required to compute the value for each pixel will vary — some will be very fast when t is exceeded in just a few iterations, while others will be slow and require the iteration limit n_{max} to be reached.

Subdividing the problem into just a few big tasks that exactly correspond to the number of processors available could mean that some blocks will finish quite rapidly relative to others. The net result will be some processors finishing early and sitting idle until those with regions of the image that iterate longer finish. So, we must find a balance between these two extremes of using N^2 tiny tasks for each pixel, and a set of P coarse grained tasks. This will allow us to keep our processors busy, increasing the utilization of the parallel machine.

The common approach to dealing with this problem is to split the problem into a set of tasks that is a reasonable multiple of P, say kP where k is an integer greater than 1. Each task would be composed of a problem of size $\frac{N^2}{kP}$. Why is this better than the case where $k = 1$? It is quite simple actually — we start by handing out P of these kP tasks to the set of processing elements that are acting as workers. As they finish, they look to see if additional tasks are pending to be completed. If there are some, they take them and start working. They only sit idle when no tasks remain to complete. The result is that when processors are lucky enough to get regions that complete quickly, they start on new work while processors working on regions requiring longer to complete are still computing.

Task granularity

So, one may ask why $k = \frac{N^2}{P}$ is not a reasonable value (corresponding to one pixel per task)? We must consider the reality that there is some amount of time required to hand a task to a worker and time to return the result of their computation back to the master. When the work associated with the task is large, these overhead times are tiny and not noticeable. As the granularity of the task becomes smaller, the overhead time approaches that of the computation. If the task is really small, then the overhead time dominates the time to perform the task. This can be quite devastating to performance. Let's see why.

Say the time to hand a task to a worker and get the result back to the master is called t_o while the time to perform the computation to determine the value of a single pixel is called t_c. In the case where we have only one worker that performs the whole computation, the total time is therefore $t_o + N^2 t_c$. Now, if we split the work into two tasks, then the total time becomes $2t_o + \frac{N^2}{2}t_c$. Why? Given that we only have one master, the act of handing work out and gathering results up is not performed in parallel, so the contribution of t_o grows with the number of workers. On the other hand, the portion working on the individual pixels does execute in parallel, so it decreases with the worker count.

So, for x tasks, we require $xt_o + \frac{N^2}{x}t_c$ time for the entire computation. As we can see, as x grows, the contribution to the overall time from t_o grows while that of t_c shrinks. At some point these two cross over, and the overhead dominates. Amdahl's law comes into play, and we see that the inherently sequential part of the algorithm (handing work out and gathering up results) dominates. This is very undesirable, as this means we are spending most of our time doing bookkeeping for the parallel implementation and very little time performing useful, productive computational work.

The practice of choosing the optimal granularity of tasks to achieve a good load balance while keeping task overheads small is an application specific question. It is good practice to, assuming it fits the problem, allow the task granularity to be tuned. This allows for experimentation with these parameters to tune the performance of the program, and also the ability to adjust the program as hardware changes, leading to changes in the task overhead and compute time required.

10.3.3   Cilk implementation

At this point, we should understand the sequential implementation of the problem. It is composed of two simple components — a main program involving two nested loops that iterate over the pixels, and a function that iterates Equation 10.1 to determine the value of a single pixel. We also have examined the issues related to decomposing the main program that can be parallelized (due to the independence of each pixel from each other) into a set of subproblems. How do we implement this in code? With Cilk, it is actually quite simple.

The `count_iterations` function remains untouched. Most of our work is related to the main program loops. The first step is to build a function that is modeled on the main program, but allows us to execute the computation for a small region (we'll call it a *tile*) of the overall problem. Each of these tiles will correspond to a task. We will define a tile as a region of the complex plane and a set of indices in the overall image that the region corresponds to. This is illustrated in Figure 10.3 for tiles corresponding to vertical strips of the image. We are faced with two issues — how to give the description of a task to a worker, and how to allow the worker to integrate the results of its computation into the overall result. For this problem, the task is simply defined as the coordinates of the points that bound the region it is working on, and the pixels that the region is to be discretized over. These can simply be passed as a set of scalar parameters to the worker function.

Getting the data back out is a subtler challenge. There are multiple ways that this can be achieved. In one case, we can assume that the worker threads will have access to the shared data structure representing the overall image. This is a safe assumption when using a shared memory threading model such as that provided by Cilk. With this shared image, the tasks can write directly to the shared data structure. Given that the tasks are independent and do not overlap, we need not worry about concurrently executing tasks conflicting with each other, so no concurrency control is required to protect this shared data. In a more general case, the workers will return a two-dimensional array representing the tile they computed, leaving the responsibility of integrating this task result into the overall result to the master. We will choose the first approach for this case study. Readers should note that the concept of *inlets*

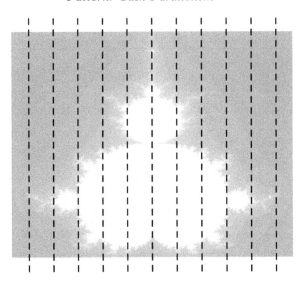

FIGURE 10.3: Mandelbrot set split into tiles for task parallel implementation.

provided by Cilk (described in Appendix C) can be used to implement the second option.

With this established, we can write a function `doTile` as shown in Listing 10.6 to process tiles on our workers.

This function is given the coordinates of the corners $U = a + bi$ and $L = c + di$. It is also given the indices of the region of the image array into which it is to store these values bounded by (ux, uy) and (lx, ly). Using these, it computes the dx and dy values and loops over the region of the image it is responsible for setting the pixels to the value provided by `count_iterations`.

Our final task is to modify the main program to not iterate over the image, but to instead generate the parameters that define the tasks that are performed by the `doTile` function. This is shown in Listing 10.7.

This program, when executed, will spawn a set of workers implementing the `doTile` function. Unlike the previous example with the genetic algorithm, the workers are not persistent. Instead, a new worker is created for each task. We rely on the underlying language and runtime to ensure that the creation of each worker is efficient by removing the responsibility of thread pooling from the programmer. We achieve load balance because the program as written spawns a set of tasks larger than the number of processors, but the underlying language executes only as many as the hardware can support. As these tasks complete, the language and runtime automatically hand off spawned tasks that are pending to processors that become free.

Listing 10.6: The `doTile` function that is executed by workers.

```
cilk void doTile(double a, double b,
                 double c, double d,
                 int ux, int uy,
                 int lx, int ly) {
  int i,j;
  double cur_a = a, cur_b = b;
  double dx, dy;

  dx = (c-a) / (double)(lx-ux);
  dy = (d-b) / (double)(ly-uy);

  for (i=ux; i < lx; i++) {
    cur_a = a + ((double)(i-ux)) * dx;
    for (j=uy; j <ly; j++) {
      cur_b = b + ((double)(j-uy)) * dy;
      I[i][j] = count_iterations(cur_a,cur_b);
    }
  }
}
```

Listing 10.7: Core of the main program for the Cilk version using **spawn** to start workers for each tile.

```
a = -2.0; b = -1.25;
c =  0.5; d =  1.25;

h_incr = (c - a) / nTiles;

for (i=0;i<nTiles;i++) {
  ta = a + ((double)i)*h_incr;
  tb = b;

  tc = a + ((double)(i+1))*h_incr;
  td = d ;

  spawn doTile(ta, tb, tc, td, i*(N/nTiles),      0,
                                (i+1)*(N/nTiles), N);
}

sync;
```

10.3.4   OpenMP implementation

OpenMP (see Appendix A) can be used to create a parallel version of the Mandelbrot program. Before we consider the program, it is useful to remind ourselves of the goals of the OpenMP language. First and foremost, the goal of OpenMP was to create a relatively simple language for a common subset of parallel algorithms. Simplicity was the goal, not generality. Second, OpenMP is designed to do as much as possible with pragma directives that are for the most part semantically neutral (at least for race-free programs). Finally, OpenMP was designed for cases where parallelism is added to an existing serial program with the hope that the parallel program will only require minor changes.

With these goals in mind, we start our consideration of the OpenMP version of the Mandelbrot program with the original serial version in Listing 10.4. A parallel OpenMP version of this program can be created by adding a single pragma before the first loop

```
#pragma omp parallel for private(j)
```

The loop control index for the first loop (i) is private by default; this is why we had to only declare the second loop's index to be private (i.e., local to each thread). This directive tells the system to create a number of threads (where we are using the default value in this case) and to split up iterations of the immediately following loop among these threads. Since we don't specify a schedule, we leave it to the compiler to choose any of the standard schedules defined in OpenMP.

Notice that this would essentially split up single pixel wide strips as a task. If you look at Figure 10.2, you see that the workload for each strip will vary widely. A better approach is to combine the loops so OpenMP defines tasks in terms of individual pixels. We do this by adding the collapse clause so our pragma reads

```
#pragma omp parallel for collapse(2)
```

This combines the following two loops into a single loop and then schedules tasks from the iterations of the combined loop. The work per task is still going to vary widely. Hence we need to consider different schedules for the loop. The major schedule options used in OpenMP are described in the OpenMP appendix. For this case where we know the effort required by each pixel will vary widely, we suspect a dynamic schedule will be needed. We switch to a dynamic schedule by changing our single pragma to

```
#pragma omp parallel for collapse(2) schedule(dynamic)
```

Each thread will complete the computing for a pixel and then return to the runtime system to pick up the next task. This leads to excessive overhead. We can change the granularity of the tasks by adding an integer parameter to the schedule clause to define a chunk size. In this case, the thread will grab "chunk" iterations each time it grabs its next set of chunks

```
#pragma omp parallel for collapse(2) schedule(dynamic, 10)
```

Notice that we have only changed one line of code in the original serial program by adding a single pragma. OpenMP has extremely low viscosity as it's almost trivial to make small changes to the algorithm.

10.3.5   Discussion

In this example, we showed how to decompose a problem into a solution based on the task parallel model where we can achieve dynamic load balancing. The Cilk language and runtime assist us in balancing the load across a fixed set of processors. We also see how the use of a shared memory can be used to integrate the results from each task into the overall global result that we wished to compute. This differs from the previous example where message passing was the mechanism for communicating tasks and the results of their computation between the master and worker.

OpenMP was particularly simple for the Mandelbrot program requiring the addition of only one semantically neutral pragma. When OpenMP is appropriate for a problem, it is hard to beat it for simplicity. This simplicity comes at a price. Notice that the programmer is restricted to only a simple canonical form of loops when using the loop workshare construct. Furthermore, with OpenMP you are restricted to the schedules selected by the language designers. The addition of explicit tasks in OpenMP 3.0 (see Appendix A) would allow us to write a Mandelbrot algorithm almost identical to the one used for Cilk. This is major expansion in the range of problems that can be addressed by OpenMP. But it doesn't completely avoid the central issue; in OpenMP we choose to trade off generality for simplicity. There will always be problems that just don't fit with OpenMP, and that's OK. One language does not need to handle all problems.

10.4   Exercises

1. A search engine may be implemented using a farm of servers, each of which contains a subset of the data that can be searched. Say this server farm has a single front-end that interacts with clients who submit queries. Describe how a query submitted to the front-end that requires searches to be executed on multiple servers in the farm can be implemented using a master-worker pattern.

2. Processing a set of images that are unrelated is a naturally task parallel activity, possibly utilizing one task per image. Consider a situation where a set of images is created from the frames of a movie, and we wish to implement an algorithm for detecting motion by identifying specific types of changes that occur between adjacent frames. Why would this motion detection algorithm not exhibit the same embarrassingly parallel structure that processing a set of unrelated images has?

3. In section 10.2.3 we point out that the static scheduling of tasks by the genetic algorithm code presented can lead to load imbalance. Modify the algorithm to use a more dynamic allocation of tasks to workers to achieve better load balancing.

4. The following C code fragment is used to multiply a pair of square matrices:

```
double A[N][ N], B[N][N], C[N][N];
for (int i=0; i<N;i++)
    for (int j=0; j<N;j++)
        for (int k=i, C[i][j]=0.0; k<N, k++)
            C[i][j] += A[i][k]*B[k][j];
```

Design a program to do this multiplication in parallel using a master-worker algorithm. How does the granularity of the tasks impact the scalability? The master-worker pattern is an unusual choice for matrix multiplication. What situation might lead to it being the right choice?

5. The OpenMP Mandelbrot program used a dynamic schedule to produce an effectively balanced load. Could a well balanced load be produced with a static schedule? What advantages would the static schedule have over the dynamic schedule?

Chapter 11

Pattern: Data Parallelism

Objectives:

- Introduce the data parallel algorithm pattern.
- Demonstrate the pattern in the context of matrix multiplication and a simple cellular automaton using array notation.
- Discuss the limitations of the pattern for general-purpose parallel programming.
- Discuss the use of task parallel constructs for approximating data parallelism.

Parallel algorithms fundamentally define tasks that execute concurrently. In some problems, the most natural way to express these tasks is in terms of the data that is modified in the course of the computation. In these problems, the data is decomposed into relatively independent components and the tasks are defined as the concurrent updates to these data components. These algorithms are known as *data parallel algorithms*.

11.1 Data parallel algorithms

The term "data parallelism" describes a wide range of parallel algorithms. At one extreme, the data is defined in terms of arrays that are distributed about the parallel system and the tasks are defined as concurrent applications of a single stream of instructions to the individual elements of the arrays. In other cases the data is decomposed into large blocks of more complex data structures and the tasks are defined as updates to these blocks with considerable variation in the computation from one block to another.

The common theme in all data parallel algorithms is an abstract index space that is defined and distributed among the computational elements within a system. Data is aligned with this index space and tasks occur as a sequence of instructions for each point in the index space.

In this chapter, we will consider the two most common classes of data parallel algorithms. In one case, there is strictly a single stream of instructions applied to each point in the index space. These algorithms are called Single Instruction Multiple Data or SIMD algorithms. This class includes vector processing. The second class is more flexible in that the instructions applied to each point in index space may be slightly different. Thinking of the index space as defining a geometry of points, these algorithms are often called Geometric Decomposition algorithms.

SIMD Parallelism

In the SIMD parallelism pattern [68], a single stream of instructions are applied concurrently to each element of a data structure. To apply this pattern, the programmer must:

- Express the problem in terms of a sequence of arrays and instructions applied across the elements of the arrays.

- Restructure the computation to remove dependencies between successive elements of the arrays so the updates within an array can safely occur in any order.

- Add instructions to clean up boundary conditions or other cases where a subset of array elements need special attention.

This approach should be familiar to programmers used to working with architectures that support vector instructions such as the vector supercomputers from Cray Research or more recently the SSE instructions found on microprocessors from Intel Corporation.

As we will see later in the chapter, the SIMD approach can reach well beyond simple vectors with instruction level support in a microprocessor. The SIMD pattern is well supported with the loop-level parallelism in OpenMP, data parallel instructions in HPF, or even careful use of the SPMD pattern in MPI.

Geometric Decomposition

Geometric Decomposition is a more general approach than that used with the SIMD pattern. The key feature of the Geometric Decomposition pattern is a large regular data structure that is decomposed into blocks or tiles. Tiles share data across their edges and update their interiors as the computation proceeds. This is a data parallel algorithm since the parallelism comes from concurrent updates to the tiles. But it is more general than the SIMD pattern since the operations applied to each tile can vary considerably.

For example, the Geometric Decomposition pattern is very common in dense linear algebra problems. In these problems, the matrices making up the problem are decomposed into a grid of tiles. The problem is defined in terms of a sequence of updates that sweep through the tiles. For each update, the local operation is applied to a tile, and then other tiles are modified as needed to keep them consistent with the updated tile. Once the full sequence of updates is completed, the computation terminates.

We will see examples of this pattern later in this chapter.

Supporting patterns

The SIMD and Geometric Decomposition patterns can be supported in a number of ways; the choice being driven in many cases by the underlying programming models available to the programmer.

The SIMD pattern is most easily supported by a programming language that is "strictly data parallel" and includes support for built in array types with operations applied "for each" element in the array. These programs consist of a single stream of instructions and hence do not require the programmer to understand multiple interacting threads and to detangle interleaving memory updates. This is a huge benefit to the programmer, but it comes at the cost of only handling parallel algorithms that can be defined in terms of a single stream of instructions.

We can relax the strict data parallel approach a bit by expressing the data parallel operations in terms of loops that are executed in parallel. This approach aligns nicely with OpenMP. The loop iterations imply an index space and the body of the loop can aggregate sets of instructions to be applied at each point in the index space. This gives the programmer more flexibility in defining the instructions applied to each element of an array, but it also can allow the instructions to vary between elements through the use of conditional logic in the loop.

At the furthest extreme is the SPMD or "Single Program Multiple Data" pattern. In this pattern, a single program is run for each processing element in a system. The programs modify their execution depending on their rank or ID within the set of processing elements. By mapping the index space for the data parallel problem onto the IDs of the concurrent multiple programs, data parallel algorithms naturally map onto the SPMD pattern. But the instructions applied by each of the multiple programs can vary greatly due to conditional logic in the program. Hence, the "data parallel" program when using the SPMD program can easily evolve into a task parallel program if the instructions applied for each point in the index space start to vary too much from one point to another.

We will see examples of each of these approaches later in this chapter.

11.2   Case study: Matrix multiplication

It is practically required of any text on parallel computing to address the classic "hello world" algorithm of parallelism — matrix multiplication. As such, you have probably encountered it at least once in the past, so we will not spend much time on it here. It will simply serve as our first realistic application of data parallel syntax to give you a quick feel for how it works with a common algorithm.

To refresh, given two matrices A and B that have dimension MxN and NxP respectively, the matrix C of dimension MxP that results from multiplying them is defined for each element of C by

$$C_{ij} = \sum_{k=1}^{N} A_{ik} * B_{kj}.$$

As we can see from examination, each element of C is the result of multiplying two vectors of length N — a row of A by the corresponding column of B. Using the notation provided by languages with rich array types, such as Fortran 90 and later or MATLAB® and Octave, we can write this very concisely.

Let's first consider a single array and the operations we can perform on it. Call this array A, and assume A is two-dimensional and square, with M rows and M columns. Without loss of generality, we can also assume that we can address the element at the ith row and jth column as A(i,j). Swapping between column major and row major ordering is simply a matter of swapping the order of the indices.

The colon, often referred to as slice or section notation, can be used for one or more dimensions to refer to a subset of the array. The notation A(:,j) would refer to the entire jth column of the array, while A(i,:) would refer to the entire ith row. Each of these slices would be one-dimensional subsets of the original array (or, more precisely, two-dimensional arrays with one dimension having an extent of 1). Clearly this will be useful given our definition of matrix multiplication. We illustrate a few different uses of array slice notation in Figure 11.1 for a two-dimensional array.

The best place to start is with the basic sequential matrix multiply algorithm, which we show in Listing 11.1. We adopt Fortran-like syntax for these examples — readers more familiar with C-like languages can simply read do-loops as for-loops, and parenthesis for array indexing as being similar to square brackets. Of course, there is no syntactic equivalent for array notation in C.

Now, if we take advantage of data parallel features we can simplify this code. We will use array notation to refer to entire rows or columns at once, and utilize data parallel intrinsic functions such as taking the sum of an array.

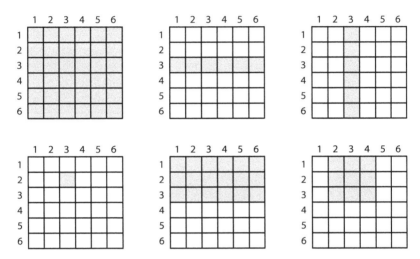

FIGURE 11.1: Six instances of slice notation. Clockwise from upper left: `A(:,:)`, `A(3,:)`, `A(:,3)`, `A(2,3)`, `A(1:3,:)`, and `A(1:3,2:4)`.

Listing 11.1: A simple sequential matrix multiply.

```
! three 2D arrays : a,b,c
real, dimension(100,100) :: a,b,c
! loop counters
integer                  :: i,j,k

do i=1,100
  do j=1,100
    c(i,j) = 0.0
    do k=1,100
      c(i,j) = c(i,j) + a(i,k) * b(k,j)
    end do
  end do
end do
```

Listing 11.2: A simple data parallel matrix multiply.

```
! three 2D arrays : a,b,c
real, dimension (100,100) :: a,b,c
! loop counters
integer                   :: i,j

do i=1,100
   do j=1,100
      c(i,j) = sum(a(i,:) * b(:,j))
   end do
end do
```

The inner loop is then easy to remove and replace with a single concise line of code. The code for the new data parallel algorithm is shown in Listing 11.2.

What do we gain by this? The code has become simpler, as the innermost loop is now gone and replaced by an implicit loop due to the use of array notation. Similarly, we no longer need to explicitly zero out the element of c that the innermost loop is computing the value of by successively adding the values of the row and column elements from a and b. The entire operation is achieved by the sum function.

As we mentioned earlier though, there is potential for a hidden increase in space required by programs that use this notation. This is due to the creation of temporaries by the compiler. In this example, the expression multiplying the ith row of a by the jth column of b in an element-wise fashion occurs as an argument to the sum function. The result of multiplying the elements of each array is a new array of 100 elements. The sum function takes the entire array as an argument and internally computes the sum of the elements to yield a single scalar. To achieve this, the compiler must emit a temporary array to store the multiplied elements and to pass into the sum function as an argument. So, while this code is shorter and uses parallel syntax provided by the programmer (instead of relying on compilation-time analysis to infer parallelism), it may end up with a higher space overhead than the original sequential code.

11.3   Case study: Cellular automaton

In this case study we examine the application of a data parallel language to the updating of a cellular automaton in time. A cellular automaton consists of a set of cells placed on a regular grid. Each cell can be in one of a finite

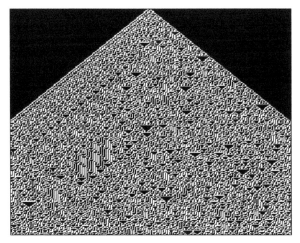

FIGURE 11.2:   An illustration of the Rule 30 1D cellular automaton over 200 time steps.

number of states, where the state of any cell c_i at time t depends on its state and the state of its neighbors at time $t - 1$. A set of rules govern the state transitions from one discrete time step to another.

A simple example is a one-dimensional cellular automaton. In this case we have an array of cells that are in one of two possible states, 0 or 1, and each cell interacts only with their immediate neighbors (one neighbor to the left and another to the right). A rule must be specified for each of the eight possible combinations of the states of a cell and that of its two neighbors. For example, one rule could be that if all three cells are initially black (in state 1), then the center cell transitions to white (state 0). Extremely complicated and intricate patterns can be generated by even this most simple of cellular automata, as exhaustively explored by S. Wolfram [94]. We show an illustration of the evolution of a 1D automaton in time in Figure 11.2. Each row in the image corresponds to the array at a point in time, with black pixels representing zeros and white representing ones. Time starts at the top and progresses down the image.

Data parallel notation is a natural way to express concurrency in an implementation of a cellular automaton. The state of the cells can be contained in an array data structure and the rules expressed as logic operations over the entire array. Fortran is chosen here to implement the case study as it has concise notation for expressing data parallelism as well as convenient transformational functions that allow the programmer to access neighbors without explicit index notation. The code segment in Listing 11.3 shows how easily one can use data parallel notation to initialize all cell values to 0 (index notation is only required for initializing the center cell value to 1).

The application of the rules for the finite automaton is shown in Listing 11.4,

Listing 11.3: Initialization of cellular automaton.

```
integer :: time = 0, finished = 100
integer :: cells(NUMCELLS), next(NUMCELLS)
integer :: left(NUMCELLS), right(NUMCELLS)

!  ... initialization of all elements to zero,
!      except for center cell

cells = 0
cells(NUMCELLS/2) = 1
```

where the rules are repeatedly applied within the time loop (only one rule is explicitly shown). A key portion of the loop is the temporary copies of the `cells` array. The **left** and **right** neighbor arrays contain the values that are to the left and the right of a given point in `cells`, respectively. The neighbor arrays are obtained by the use of the Fortran function `cshift` that applies cyclic boundary conditions . The text within the **where** block sets the elements of the temporary array **next** to 0, *only where* the element in the `cells` array is 1, *and* the neighboring values are also 1.

This example shows the concise way in which data parallel languages can be used to express some algorithms. They work best for problems requiring identical operations to be applied to all elements of an array or other set-based data structure. Some languages allow operations to be applied on neighboring elements as well. The data parallel notation provides enough information to the compiler so that it can emit code that can be run concurrently, because a data parallel operation can be applied to all the elements of an array in any order. However the data parallel notation only provides semantic hints to the compiler; the degree of concurrency you obtain with an actual compiler will vary.

11.4   Limitations of SIMD data parallel programming

Data parallel programming is very attractive for at least two reasons. First, it is geared towards concise expression of operations that occur on large blocks of data. Explicitly programmed loops disappear for operations that are repeated over an array of elements, replaced with a single statement written in terms of the entire array (or array subset). Second, many hardware acceleration technologies exist that are data parallel by nature. Recall that data parallelism was the driving force behind the design and construction of

Listing 11.4: Time loop for cellular automaton implementing the one-dimensional Rule 30.

```
do while (time < finished)

    ! ... make working copies

    next  = cells
    right = cshift(cells, 1)
    left  = cshift(cells, -1)

    ! ... apply rules

    where (left  == 1 .and. &
           cells == 1 .and. right == 1) next = 0
    where (left  == 1 .and. &
           cells == 1 .and. right == 0) next = 0
    where (left  == 1 .and. &
           cells == 0 .and. right == 1) next = 0
    where (left  == 1 .and. &
           cells == 0 .and. right == 0) next = 1
    where (left  == 0 .and. &
           cells == 1 .and. right == 1) next = 1
    where (left  == 0 .and. &
           cells == 1 .and. right == 0) next = 1
    where (left  == 0 .and. &
           cells == 0 .and. right == 1) next = 1
    where (left  == 0 .and. &
           cells == 0 .and. right == 0) next = 0

    ! ... obtain results from working copy

    cells = next
    time  = time + 1

end do
```

many early vector processing supercomputers. The multimedia extensions that arrived in desktop processors in the 1990s and graphics processing units in recent years are all not-very-distant relatives of the same technology that was pioneered in vector computers. It has long been known that extremely fast hardware can be constructed to execute code written in a data parallel fashion.

Unfortunately, as was discovered in the early days of vector supercomputing, there are limits to the flexibility of vector processing. Some algorithms simply cannot be easily recast in a vectorized form such that data parallel hardware can execute it efficiently. Most often this is due to dependencies in the code that restrict the amount of available parallelism. For example, complex loops with dependencies across iterations can be quite difficult (if not impossible) to turn into a vectorized form. More often than not, one must go back to the drawing board and design a new algorithm to solve the problem that has a higher amount of parallelism present.

11.5 Beyond SIMD

Typical algorithms that work on large data structures such as arrays are often composed of small operations that work on small regions of the structure which are iteratively applied to each element of the structure. For example, in image processing a common operation is to apply some function to the intensity of each pixel to perform contrast enhancement. A rudimentary form of image segmentation (partitioning the intensities into a small set of groups) is implemented by assigning a value to a pixel corresponding to it falling within a set of specified intensity ranges. The most basic technique is known as *thresholding*, and simply splits the intensity range into two pieces, where pixels in one range get a value of zero and the remainder get a value of one. We see some simple C code for this in Listing 11.5

The algorithm itself has a data parallel structure to it — a single operation is performed for each element of the array to generate a new array containing the results of applying that operation. We could write this simple function as:

```
int threshold(int x, int t) {
  if (x < t) return 0;
  else return 1;
}
```

The algorithm then becomes nothing more than applying this function to the whole array. In languages that support elemental functions, this would be handled directly by the language making the loops over the array implicit

Listing 11.5: Binary thresholding.

```
int x[m][n];
int i,j;

for (i=0; i<m; i++)
  for (j=0; j<n; j++)
    if (x[i][j] < t)
      x[i][j] = 0;
    else
      x[i][j] = 1;
```

in the function invocation. Another approach is to use languages that provide syntax for invoking these small functions (often called *kernels*) over the whole array using specialized syntax. New language extensions such as CUDA and OpenCL provide tools for programmers to do these types of operations. Unfortunately, these languages are nearly as concise as those that provide elemental functions and often require the programmer to do many low-level operations to move data around before and after the kernel is executed in parallel over the arrays. For example, in CUDA, the programmer must allocate memory by hand on both the host (the CPU side) and the computational device (a graphics processing unit), and explicitly copy the data to the device before starting the kernel, and finish by explicitly copying the results from the device when it completes. So, while CUDA provides a data parallel model to the programmer, the degree to which the language helps abstract above the operation is limited due to these exposed low-level operations.

11.5.1 Approximating data parallelism with tasks

As we have discussed previously, the use of array notation to express operations over entire arrays allows the computations on individual elements to be executed concurrently. Sequential loops become single array expressions, so adopting MATLAB notation, the loop

```
for i=1:n
  x(i) = a(i) * b(i);
```

can be represented as a single operation:

```
x = a .* b;
```

The key is that by removing the explicit loop and allowing the language interpreter or compiler to choose an efficient way to implement the whole-array operation, parallelism can be exploited. We can achieve a similar extraction of parallelism from the explicit loop by using task-parallel constructs such as those provided by OpenMP to approximate what we achieve with array notation in languages that do not provide it (such as C or C++). Even if we build array classes in C++ to provide a higher-level abstraction closer to array notation to a programmer, we would still need to parallelize the loops internally within the array class implementation. A likely choice for that would be OpenMP annotations. For example, with the loop above, we can use the parallel `for`-loop provided by OpenMP to execute the loop iterations in parallel.

```
#pragma omp parallel for private(i)
for (i=0; i<n; i++)
  x[i] = a[i] * b[i];
```

The result is that a set of tasks will be created by the compiler for each loop iteration, and the updates to `x[]` would execute in parallel. We intentionally are calling this an "approximation" of data parallelism because in reality, we are using the work-sharing constructs that OpenMP provides to create a set of tasks. With parallel loops such as this though, the tasks are for a very specific role — loop iterations. The net effect to the programmer is similar to data parallel array notation though — the update to the array elements executes in parallel. While not as syntactically concise, this is a step in the right direction for programmers using languages that lack data parallel facilities in the language but still wish to adopt a data parallel way of thinking about their programs.

11.6   Geometric Decomposition

Geometric Decomposition is a common pattern in parallel programming. The high-level idea is to define a geometry that extends across the problem. This can be a mesh of points spanning an image, the elements of a matrix, or any number of structures that can be thought of in terms of a spatial coverage of points. In this pattern, the geometry is decomposed into tiles or blocks. The concurrency is expressed in terms of parallel updates to these blocks.

As an example of Geometric Decomposition, let's return to matrix multiplication (Listing 11.1). We can decompose the problem geometrically in terms of the columns of the matrix C by assigning iterations of the first loop to different threads. We can do this with OpenMP by adding a single directive before the first loop

```
!$ OMP parallel do private(j,k)
```

Note the use of Fortran syntax with the `do` loop construct instead of the `for` construct from C and the structured comment "`!$ OMP`" instead of "`pragma omp`". Otherwise, OpenMP is basically the same for Fortran and C.

As suggested earlier, the Geometric Decomposition view of this problem comes from thinking of the problem as assigning columns of C to different threads. The parallelism is expressed in terms of operations that can be applied independently for each column of the C matrix.

This raises a subtle and potentially confusing point. By defining the concurrency in terms of parallel updates to blocks of data, this is a data parallel pattern. But we can think of the same code in terms of the tasks executed by the threads and hence consider this as a type of task parallelism. Which of these views is correct? The answer is both of them are simultaneously correct; it all depends on which point of view is more convenient for the programmer. It's a human factor and since human language is fuzzy, we end up with fuzziness in how we conceive of our patterns. We are using patterns to capture the way people think about parallelism. Since people think about parallelism in "fuzzy" ways, we shouldn't be surprised that we will find contextually vague features in how patterns are applied.

11.7 Exercises

1. Given an array X of N numbers, the mean of the array is defined as $\mu = \sum_{i=1}^{N} \frac{x_i}{N}$. Write this computation in data parallel notation.

2. Given an array X of N numbers, the variance of an array is defined as $\sigma^2 = \sum_{i=1}^{N} \frac{(x_i - \mu)^2}{N}$. Write this computation in data parallel notation.

3. In the chapter introducing array notation, we saw that one can approximate the derivative of an array of numbers using the operation $diff(:) = X(2 : N) - X(1 : N - 1)$. Write a similar expression to compute the second derivative of the array using array notation.

4. Consider matrix multiplication (section 11.2) and the program from exercise 4 of the last chapter. Implement the program using OpenMP or Cilk (see the appendices). How do you expect the performance of these variants to compare to the version from section 11.2 (array notation)?

5. Many scientific algorithms have a dependency on neighbors like that shown in the cellular automaton case study. For example, the temperature along a steel rod can be approximated by repeatedly updating the contents of a given cell with the the average of that cell's value and its two immediate neighbors. Write this computation in data parallel notation.

Chapter 12

Pattern: Recursive Algorithms

Objectives:

- Review the concept of recursion as a general algorithm pattern.
- Demonstrate recursion used to implement a parallel version of the classical mergesort algorithm and a simple Sudoku puzzle solver using Cilk.

Agenda parallelism, thinking of a problem directly in terms of the tasks that must be carried out, is an extremely important class of parallel algorithms. Tasks can be generated in many different ways. In Chapter 10 we discussed algorithms where the tasks could be generated statically or through an iterative control structure. In this chapter, we consider a second important class of task parallel algorithms, problems for which the tasks are generated recursively.

Recursion is one of the fundamental algorithmic techniques in computer science. The basic concept of recursion is that of a function that calls itself. The evaluation of the recursive function is created by composing together the results of the calls to itself, using successive refinements of its input data at each step. This refinement can range from dividing the data into smaller pieces, traversing a self-similar data structure like a linked list, or performing a numerical computation on the inputs. Finally, the function must have a condition to stop the repeated process of calling itself so that the calls can return and unwind to the original caller with the final result. The classic example is that of computing the Fibonacci numbers. By definition, the nth Fibonacci number is computed with the following recurrence relation (n must be a non-negative integer).

$$fib(0) = 1$$
$$fib(1) = 1$$
$$fib(n) = fib(n-1) + fib(n-2)$$

The recursive nature of the problem appears in the third rule, in which the nth Fibonacci number is defined in terms of the $(n-1)$th and $(n-2)$th Fibonacci numbers. Familiar programming constructs such as the `for`-loop can be implemented recursively also. Interestingly, a common optimization made by compilers is to take recursive functions that have a particular structure (called *tail recursive*) and turn the recursion into iteration. This transformation eliminates the overhead required for function calls and allows powerful loop optimizations to be applied to transformed code that would otherwise not apply to the recursive version.

Recursion is a powerful programming technique when the problem being implemented has a naturally recursive structure. Problems of this form appear repeatedly in practice.

In this chapter we will look at how to implement algorithms with a recursive structure. We will choose two different families of languages to draw from. First, we will discuss recursive parallelism in the context of the declarative language Erlang. This is natural, as declarative languages (most often functional languages) are often the prime example of languages and compilers that thrive on recursion. Second, we will discuss Cilk, which is a derivative of the ANSI C language. Cilk is of interest because its runtime scheduling of threads across processors was built to be able to efficiently and effectively deal with the large number of threads that may be generated when threads are spawned recursively. Furthermore, it has a clever mechanism built into the language to allow dynamic cancellation of whole subtrees within the recursive call tree. This is particularly interesting when the recursive algorithm in question is speculatively exploring a solution space. It would be premature to go into more detail now, so we'll leave that interesting feature for later in the chapter.

The reader is advised to refer to Appendices B and C for brief references for the Erlang and Cilk languages respectively.

12.1   Recursion concepts

Let's briefly remind ourselves what recursion is, why it is useful, and what ramifications it has on a language and compiler in implementing it. Mathematically, a recursive function is one in which the result it produces is defined in terms of the function itself. The basis of this is mathematical induction.

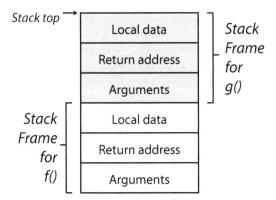

FIGURE 12.1: Basic stack discipline for a function f() calling a function g().

Induction requires the definition of some primitive base cases on which other cases are built. For example, consider a simple factorial function that computes the product of all integers from 1 up to n. A recursive definition of the factorial function $f(n) = n!$ would simply be for $n \geq 1$

$$f(1) = 1$$
$$f(n) = n * f(n-1).$$

The base case is $f(1)$, and $f(n)$ defines the general case that is based on a recursive definition in which $f()$ is applied to $n - 1$. These two cases can be used to prove properties of the function through mathematical induction, and their definition corresponds to the implementation in computer code as a recursive function. The base case provides the stopping condition to end recursive calls and begin the process of unwinding to accumulate up the final answer. The second case defines the recursive calls ($f(n-1)$), and the work performed as the calls return to stitch the final result together (in this case, multiplication by the parameter n).

The implementation of a recursive function is very similar (if not identical) to that of a regular function. In most languages, a function invocation is a simple process of:

1. Pushing the state of the caller, such as processor registers, onto a stack.

2. Pushing function arguments onto the stack (either values or references, depending on the language).

3. Pushing the return address to jump to when the callee returns.

Once the stack has been properly set up for the function call, the caller jumps to the address of the callee and the called function begins to execute.

Any function invocations it makes will follow the same pattern. Figure 12.1 illustrates the basic structure of the stack with the stack frames corresponding to a function f() calling a function g(). Functions are also able to use the stack as a location for storing local variables. When the function completes execution and returns to the caller, a similar simple stack operation occurs:

1. Pushing return values onto the stack.

2. Jumping to the return address pushed onto the stack by the caller.

This process of pushing and popping state onto the stack corresponding to function invocations is known as a *stack discipline*, and has become largely standardized over time. The exact details for how the stack is implemented for a given language and platform requires understanding the specific compiler, operating system, and instruction set.

In its most basic form, recursive functions use the standard stack discipline for invocation. In fact, the existence of a well defined stack discipline makes recursion almost automatic for a language. Older languages, such as the early versions of FORTRAN (before the case was changed to Fortran) did not support recursion, and more recent versions do with the caveat that recursive functions be explicitly labeled as recursive. This is due to the lack of a well defined stack discipline in early days of the language, and subsequent work on compiler optimizations that can take advantage of the assumption that recursion is *not* present unless otherwise stated.

Recursion is implemented using the stack to ensure that the state of each invocation of the recursive function is preserved and restored in the order defined by the function calls and returns during execution. To understand what a recursive function is actually accomplishing though, we should look at the graph that results from the parent and child relationship between function callers and callees. This is often referred to as the *call graph*.

The call graph is a directed acyclic graph in which the vertices are function invocations, and the directed edges originate from the caller and terminate at the callee. The call graph for a recursive function is particularly interesting, especially with respect to where we are heading with this discussion, which is the exploitation of parallelism to speed up algorithms based on a recursive pattern. Consider the snippet of code in Listing 12.1 that implements a simple function to compute Fibonacci numbers. Note that the code is written in this manner for illustrative purposes only — in practice, memoization would be used for algorithms of this form to avoid recomputing values repeatedly and would perform significantly faster for large values of n.

The call graph that results from an invocation of this function for $n = 4$ is shown in Figure 12.2. What is interesting here? An examination of the call graph shows that the invocation of fib(4) results in two subtrees representing the call graphs for fib(3) and fib(2). In a sequential computer, only one of these would be taken at a time, so the full subtree of fib(3) would execute

Listing 12.1: A simple Fibonacci number function.

```
int fib(int n) {
   if (n==1 || n==0)
      return 1;
   else
      return fib(n-1) + fib(n-2);
}
```

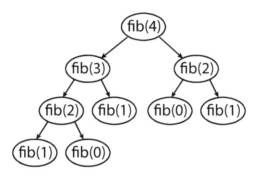

FIGURE 12.2: A call graph for an invocation of `fib(4)`.

to completion before `fib(2)` could begin. This serialization of the tree would repeat in further subtrees for each recursive call.

There is something interesting occurring here. The recursive function is defined *as if* both subtrees executed at the same time. The fact that they do not is simply a consequence of the implementation platform. Given concurrent hardware, it is entirely feasible to allow different processing elements to execute these separate subtrees of the call graph. Clearly, used properly, this division of work across processors based on the recursive structure of the algorithm could be of great benefit to performance. This is the motivating reason for pursuing recursion as a source of parallelism.

Recursion and side-effects

Side-effects caused by function invocations complicate using recursion to identify concurrency. If a recursive function has side-effects then a dependency may exist between invocations of the function. These are apparent if a correct execution of the recursive function requires a specific order of execution for the recursive call graph. To properly use recursion to exploit concurrency, one must ensure that side-effects performed by recursive calls are well understood and controlled. Failure to do so can have consequences with respect to the correctness of an algorithm in a concurrent execution environment.

12.1.1   Recursion and concurrency

Why did we begin this recursion refresher with a discussion of the stack discipline? Recursion, as with all function invocations, is intimately tied to the stack of activation records. How does this work in the context of recursive calls that execute concurrently? If these concurrent functions themselves execute recursive (or non-recursive) functions, where do they go on the stack? How can a stack discipline be maintained? In practice, two approaches are encountered. The first is based on spawning of new threads for each call. The second is based on redefining how the stack is structured.

In the first case, a new thread of execution is created for each recursive call. This approach is the easiest to implement given a thread implementation for a given programming language. By definition, threads are essentially very light weight processes, and as such, contain their own stack. This means that invocations of recursive functions that are encapsulated within threads are not actually pushed onto the stack of the caller, but reside within the independent threads of execution. Parameters are passed in as arguments to the thread, and return of control to the caller is based on the caller blocking on the child threads until their completion. This is the simplest (and, as of now, most widespread) method for implementing function calls concurrently, as it requires no modification to the underlying implementation of the language such as the function invocation stack.

In the second case, we redefine how the stack works to take into account the requirements of concurrency. This is the approach taken in the Cilk language. In Cilk, functions intended to execute concurrently with their caller (annotated as Cilk subroutines) are not explicitly encapsulated within threads, but are called just like any other regular function. Within these Cilk subroutines, non-Cilk functions are perfectly legitimate callees, expecting an activation stack that behaves as they would usually encounter. Cilk accomplishes this by not forcing the stack for each function invocation to be encapsulated within separate threads, but by providing a new type of stack known as a *cactus stack*.

A cactus stack is a parallel stack implemented as a tree data structure with child nodes containing pointers to their parent nodes. As each Cilk procedure is spawned, it is provided a reference to the current call stack with its own stack frame pushed on the top. An example is shown in Figure 12.3. This example shows a view of the stacks representing the right half of the call graph of the recursive Fibonacci invocation shown earlier in Figure 12.2. This figure shows the views of the stack as seen by each procedure, with the stack growing downward. The boxes represent stack frames (with the name of the parent procedure provided at the top of each of the stack columns).

In addition to the stack discipline used to manage function invocations, we also must consider another issue that arises when using concurrency in recursive algorithms. What happens if the recursive function invocations modify some shared state? If this occurs, we will need to ensure that data is protected

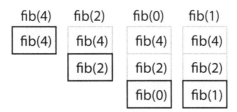

FIGURE 12.3: Partial view of the call stacks for an invocation of `fib(4)` corresponding to the right half of the call graph shown in Figure 12.2. The active stack frames are highlighted.

via concurrency control primitives to prevent these concurrent operations from colliding with each other. This need for concurrency control is no different than other concurrent programming contexts.

An issue that may be subtler to notice when turning a sequential recursive algorithm into a concurrent one is whether or not the algorithm relies on the order of recursive child calls being consistent. This only arises when a function makes multiple recursive calls, such as during the traversal of a binary tree. In that case, a function may recursively call itself for the left subtree and then the right subtree. In some computations it is possible that the order of traversal (such as left before right subtrees) must be maintained for the algorithm to be correct. When these calls are executed concurrently, all bets are off with respect to their order of execution unless the programmer puts in place concurrency control mechanisms to enforce the needed order.

12.1.2 Recursion and the divide and conquer pattern

The divide and conquer approach to algorithm design and problem solving is very common. The basic premise of this approach is that a difficult problem can be solved by dividing it into simpler subproblems that can be solved independently, with their results being recombined to form the solution to the larger complex problem. There is a natural fit to concurrent computing here, as the independence of the subproblems implies that they can be solved at the same time without conflict.

Divide and conquer approaches to problem solving appear in many places, and are frequently defined recursively. For example, when sorting a list of values, one can do so with the classic mergesort algorithm, in which the list is divided into halves and the halves sorted independently, with the resulting sorted sublists being merged efficiently back together to form the whole sorted list. This is applied recursively, such that each half is then divided further into halves and merged, down to the point where the divided sublists eventually reach lengths of one or two elements, at which time sorting is trivial. The traditional use of tree structures is also based on this notion of divide and conquer, not only with respect to algorithms that work on them, but in the data representation.

It is not true that divide and conquer algorithms are all recursive, nor are all recursive algorithms a form of the divide and conquer design pattern. In fact, any parallel algorithm is really a realization of the divide and conquer pattern. Work is divided up amongst a set of processing elements who all work together to achieve the end goal and "conquer" the problem. We make special mention here of the divide and conquer pattern simply because many common algorithms based on it are defined using a recursive formulation.

12.2   Case study: Sorting

The classic example of recursion in practice is that of sorting a large set of values with an algorithm such as mergesort or quicksort. The basic premise of these algorithms is as follows. Given a list of elements X of length N, they can be sorted by dividing the set into two subsets $X_{1..M}$ and $X_{M+1..N}$. Mergesort typically splits the lists in half, where $M = \frac{N}{2}$. Each half list can then be sorted independently, with a simple merge operation occurring that recombines the subproblem results back together to form the solution to the larger problem.

Let us remind ourselves of how the traditional mergesort algorithm is typically stated.

- Divide the unsorted sequence into two sublists of roughly equal size.

- Recursively call mergesort on each of the two sublists.

- Merge the two sublists (now sorted) into a sorted sequence of items.

As we can see, at each stage (other than the base case), there are two recursive invocations of mergesort on the two halves of the input list. These are performed independently of each other, only coming together after they have completed in order to be merged to form the full sorted list. Given that these halves are operated on completely independently before the merge step, it should be clear that they can be executed in parallel — no data dependency exists between them. The only requirement is that both sublist sorts complete before the merge phase is entered that puts them back together.

The implementation of the mergesort algorithm in C is shown in Listing 12.2. The input list X is split in half around the pivot point n/2 and each half is given to the **mergesort** function sequentially; the lower half sorted

Listing 12.2: Mergesort algorithm.

```
void mergesort(int *X, int n, int *tmp) {
  if (n < 2) return;

  /* recursively sort each half of list */
  mergesort(X,n/2,tmp);
  mergesort(X+(n/2),n-(n/2),tmp);

  /* merge sorted halves into sorted list */
  merge(X,n,tmp);
}
```

first and the upper half sorted next. The recursion completes when the terminating condition n < 2 is reached. Finally, the sorted sublists are merged together by the call to merge. Note the array tmp that is used for temporary storage with the same length n as does X.

Listing 12.3 shows the merge function. This function first gathers items from the two sublists in sequential order by copying into tmp in lines 6–14. When copying from one of the sublists has completed, the rest of the other sublist is copied into tmp in lines 15-18 or 19-22. Finally the sorted list is copied back to X.

Because the two recursive calls to mergesort in Listing 12.2 operate on entirely separate portions of the input list, there is independence between the input data in the two function calls and the two calls can be safely run concurrently. The beauty of recursive, divide-and-conquer algorithms like mergesort is that each division of the task can often be run concurrently. Listing 12.4 shows an implementation of mergesort in Cilk. As you can see, the Cilk implementation is nearly identical to the C version; the only differences being the use of cilk keyword to declare the function in line 1, the spawning of the two mergesort calls in lines 5 and 6, and the call to the synchronization primitive sync in line 9. The synchronization ensures that each thread completes before the two sublists are merged together in line 12. Otherwise, the merge process would likely take place while before the sublists are fully sorted, clearly an error in implementation.

When executed on very large arrays the simple Cilk implementation we show here outperforms a sequential implementation. For example, we performed an experiment where a 32,000,000 element array populated with random integers was sorted sequentially with the original C version and in parallel with Cilk using only the modifications we made here. On a 2.4GHz quad-core Xeon, we observed a time of approximately 5.5 seconds to sort the list using all four cores by the Cilk code, but nearly 8 seconds using the sequential version of the code. While not an optimal speedup of 4x when using four times

Listing 12.3: Merge function used by both C and Cilk implementations.

```
1  void merge(int *X, int n, int *tmp) {
2    int i = 0;
3    int j = n/2;
4    int ti = 0;
5
6    while (i<n/2 && j<n) {
7      if (X[i] < X[j]) {
8        tmp[ti] = X[i];
9        ti++; i++;
10     } else {
11       tmp[ti] = X[j];
12       ti++; j++;
13     }
14   }
15   while (i<n/2) { /* finish up lower half */
16     tmp[ti] = X[i];
17     ti++; i++;
18   }
19   while (j<n) {    /* finish up upper half */
20     tmp[ti] = X[j];
21     ti++; j++;
22   }
23
24   memcpy(X,tmp,n*sizeof(int));
25 }
```

Listing 12.4: Cilk mergesort based on Listing 12.2.

```
1  cilk void mergesort(int *X, int n, int *tmp) {
2    if (n < 2) return;
3
4    /* concurrently sort each half of list */
5    spawn mergesort(X,n/2,tmp);
6    spawn mergesort(X+(n/2),n-(n/2),tmp);
7
8    /* wait for both threads to finish */
9    sync;
10
11   /* merge sorted halves into sorted list */
12   merge(X,n,tmp);
13 }
```

the core count, the fact that the such simple modifications to the code can yield noticeable speedups is quite attractive.

Recursive algorithms such as the one used with Cilk in Listing 12.4 are very important. As parallel programming expands its reach into general purpose programming, the importance of these algorithms will grow. OpenMP as originally conceived did not handle recursive algorithms very well. This changed with OpenMP 3.0; a major upgrade of the language. As described in Appendix A, OpenMP was redesigned to incorporate explicit tasks into the OpenMP programming model. This lets a programmer create recursive algorithms in a style familiar to Cilk programmers within OpenMP.

Even with explicit tasks, the only way to create threads in OpenMP is to utilize a parallel construct. Hence we need to assume that the code calling the merge function is contained within a parallel region. We then can express the same algorithm as used with Cilk in Listing 12.4 with explicit tasks in OpenMP 3.0. We show this code in Listing 12.5. If you understood the Cilk version of the program, you can directly infer what the OpenMP program is doing. A task is created and added to a task pool. Available threads grab tasks and execute them until the thread pools are empty. The `firstprivate` clause is needed to protect the values of the variables X, n and `tmp` so the created tasks see their values as they were when the task was created. Finally, the `taskwait` pragma is exactly the same as the Cilk `sync`; it causes the parent thread to wait until its children return. This is just a quick summary to show how Cilk functionality for recursive algorithms can be captured with OpenMP 3.0. There are many additional low-level details to support this major addition to OpenMP. To learn more, consult the summary in Appendix A or download the OpenMP specification.

12.3   Case study: Sudoku

Sudoku[1] is a popular puzzle that involves filling in a grid with numbers based on a set of constraints. A 4×4 Sudoku grid is shown in Figure 12.4. The constraints are based on rows, columns, and regions. Regions are defined as smaller squares within the grid. In the 4×4 case, these would be the 2×2 blocks formed by taking the board and splitting it in half both horizontally and vertically. In conventional Sudoku puzzles found in newspapers, we are given a 9×9 grid with nine 3×3 regions. A given number must only appear once in each constraint dimension (row, column, region). The possible numbers that can be placed in the squares for an $n \times n$ board with n regions are 1 through n. For example, the solution to the puzzle in Figure 12.4 at row 3, column

---

[1]Sudoku is a Japanese word meaning "the numbers must be single."

Listing 12.5: OpenMP mergesort based on Listing 12.2.

```
1  void mmergesort (int *X, int n, int *tmp){
2    if (n < 2) return;
3
4    /* concurrently sort each half of list */
5    #pragma omp task firstprivate(X, n, tmp)
6    mmergesort(X, n/2, tmp);
7
8    #pragma omp task firstprivate(X, n, tmp)
9    mmergesort(X+(n/2), n-(n/2),tmp);
10
11   /* wait for both threads to finish */
12   #pragma omp taskwait
13
14   /* merge sorted halves into sorted list */
15   mmerge(x,n,tmp);
16 }
```

4 is 1, because all of the rest of the numbers have been taken in column 4. Once this position has been filled in (with the number 1), the only available numbers for the lower-right 2×2 region are 2 and 4.

A simple strategy is just to guess the answer for each location in the grid and halt when the correct solution (based on the constraints) is found. However, assuming m open positions in an $n \times n$ grid, a simple guessing strategy involves (ignoring the constraints) n^m separate possible solutions (n possible numbers in each of the m open positions). For the $m = 11$ open positions in Figure 12.4, 4^{11} is approximately 4 million.

Given enough avenues for concurrency p, this may not be prohibitively expensive. However for a typical 9×9 grid there are over 10^{47} possible solutions (again ignoring the constraints) and guessing a solution would take on the order of $200/p$ days. Sometimes a simple (though computational inefficient) algorithm is appropriate given sufficient computational resources, especially when the program will be run only a few times and programmer time is taken into consideration. Even when a task is run repeatedly in a production regime (such as a weather prediction code) it may be practical to *start* with a simple algorithm and refine it later. This is the approach we take with the Sudoku solver (in fact a simple refinement of the simple guessing strategy is possible and is left as an exercise for the reader).

The guessing strategy is naturally recursive. The function solve starts with a partial solution to the puzzle as input, guesses k numbers for the first open position, calling solve k times (one for each of the k guesses) with the first open position in the puzzle filled in with the number guessed. If there

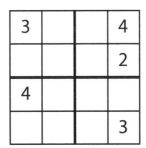

FIGURE 12.4:   A 4 × 4 Sudoku puzzle with 11 open positions.

are no more open positions, `solve` checks the solution and returns true if the current solution is correct or false otherwise.

The tasks in a concurrent solution using the guessing strategy are the separate calls to `solve`. Having identified the tasks, we next must decide how data is to be shared between the tasks. This is usually the most difficult and dangerous part of parallel programming as there may be sharing of data structures between concurrent tasks. However, in this instance, if a new copy of the solution space is provided to each task, the task can work (fill in empty positions in the grid with guesses) on its own data independently of all the other tasks.

The simple guessing strategy naturally maps onto the Cilk language, where each call to solve can be spawned as a separate task. Relevant portions of the `solve` function implemented in Cilk are shown in Listings 12.6–12.8. Note the calls to the C library functions from Cilk (e.g., `check_solution`). These C functions are provided in `lib_sudoku` on the book's Web page as a convenience to the reader.

The first part of this implementation in Listing 12.6 checks to see if the last position in the puzzle has already been encountered, and if so, `solve` returns whether or not the solution is correct at line 9. If the current location already contains a number (given by the original puzzle), the implementation moves on to the next puzzle position by calling `solve` at line 15.

A key to successfully using recursion with Cilk is to ensure that each thread has its own copy of the puzzle to work on; this is provided by the call to the Sudoku library function `copy_grid` at line 6 of Listing 12.7. As its name implies, this function provides a newly minted copy of the current working copy, `grid`. In Listing 12.7, each of the `numNumbers` successive guesses are tried concurrently by spawning a call to `solve` at line 8 with the newly guessed number `k+1` inserted in the new grid `mygrid[k]`. The solution for each of the guesses is stored in the vector `solution`.

Listing 12.6: Sudoku guessing algorithm implemented in Cilk, part *a*.

```
1  cilk int solve(int size, int *grid, int loc) {
2    int i, k, solved, solution[MAX_NUM];
3    int *myGrid[MAX_NUM];
4    int numNumbers = size*size;
5    int gridLength = numNumbers*numNumbers;
6
7    if (loc == gridLength) {
8      /* we've reached maximum depth */
9        return check_solution(size, grid);
10   }
11
12   /* if this node has a solution at this location */
13   /* (given by the puzzle), move to next location */
14   if (grid[loc] != 0) {
15     solved = spawn solve(size, grid, loc+1);
16     return solved;
17   }
18   ...
```

Listing 12.7: Sudoku guessing algorithm, part *b*.

```
1    ...
2    /* try each number (unique to row,column,square) */
3    numGrids = 0;
4    for (k = 0; k < MAX_NUM; k++) {
5      /* need new grid to work with */
6      myGrid[k] = copy_grid(size, g);
7      myGrid[k][loc] = k+1;
8      solution[k] = spawn solve(size, myGrid[k],
9                                loc+1);
10     numGrids += 1;
11   }
12
13   /* wait for all children to complete */
14   sync;
15   ...
```

Listing 12.8: Sudoku guessing algorithm, part *c*.

```
1    ...
2    /* check to see if there is a solution */
3    solved = 0;
4    for (k = 0; k < numNumbers; k++) {
5      if (solution[k] == 1) {
6        int n;
7        /* success, copy solution to parent's grid */
8        for (n = loc; n < gridLength; n++) {
9          grid[n] = myGrid[k][n];
10       }
11       solved = 1;
12     }
13     free(myGrid[k]);
14   }
15
16   return solved;
17 }
```

At line 14 the implementation waits until all of the children have completed. This is necessary to ensure that the solution vector has been filled in by all of the children. At this point we check whether any of the children have found a correct solution at lines 4–14 in Listing 12.8. If a correct solution has been found, the original grid from the parent is filled in with the solution found by one of the children at 8-10.

12.4   Exercises

1. Dynamic programming is an important computational pattern used in a wide range of problems (e.g., the Smith Waterman algorithm for genome sequence matching). A key feature of dynamic programming problems is recursion with an overlapping problem domain so subproblems appear multiple times. By saving earlier solutions to the subproblems, recomputing solutions to these problems can be avoided and the computation is accelerated. The process of saving subproblem solutions for later reuse is called "memoization." Take the Fibonacci sequence problem from section 12.1 and show how to convert this to a dynamic programming problem by adding memoization.

2. Improve the efficiency of the Cilk Sudoku solver by only recursively searching for solutions with numbers that fit the constraints known for the puzzle at each step; this should speed up the solution dramatically as the constraints will reduce the number of attempted solutions.

3. Adapt the Sudoku solver to run serially using only C. Measure the time it takes to reach a solution and plot how this time compares with the Cilk version on various numbers of processors available to you. Discuss these measurements. Are two processors faster than the serial version? If not, why not? Can you find a way to improve the parallel performance?

4. Using the cognitive dimensions from Chapter 5, compare and contrast the Cilk and OpenMP implementations of the mergesort program.

5. Recursion can be a natural way to express an algorithm, but it can be challenging to support efficiently at runtime on a parallel system. Consider carefully how the mergesort program would execute for extremely large arrays. What are some ways a recursive algorithm might run into trouble while executing this program? Can you suggest some runtime optimization to mitigate these problems?

Chapter 13

Pattern: Pipelined Algorithms

Objectives:

- Introduce the pipeline pattern for algorithm design and discuss how it reveals concurrency in the solution to a problem.
- Demonstrate the pipeline pattern as implemented using message passing in the Erlang language.
- Describe the pipeline-based structure of a simulation of the visual cortex in the brain.

In previous chapters, we discussed agenda parallelism (task and recursive parallelism patterns) and results parallelism (data parallelism patterns). In this chapter, we address our third strategy for exploiting concurrency in a parallel application — *specialization*, or *specialist parallelism*.

The central idea in specialist parallelism is to express the problem in terms of a set of distinct specialized tasks between which data flows in-order between the tasks to create a processing pipeline. All tasks must execute for each block of data input into the system, but since the specialized tasks run at the same time, concurrency is exposed.

The pipeline is a common abstraction in computing. In modern CPUs, pipelining is used inside the arithmetic units to extract instruction level parallelism from serial programs. Specialized hardware for digital signal processing and graphics processing make heavy use of pipelining. Outside computing, pipelining is the central idea behind the assembly line used in manufacturing to increase efficiency and throughput.

When using a pipeline, a problem is split into a sequence of smaller tasks. Each of these smaller tasks is known as a *stage* of the pipeline. Data flows through the pipeline, being input at one end and output from the other. Each stage passes its output as input to the next stage in line. We think about the rate at which data passes through the pipeline in terms of time steps — one time step being the amount of time required by a stage to process one input and produce one output. Once the pipeline is full and enough data has entered the sequence such that each stage is performing work, a new result is produced out of the last stage at each time step.

The easiest example for visualizing a pipeline comes from the familiar assembly line used in manufacturing. Consider an assembly line to build an automobile. The assembly line is composed of a sequence of stations (pipeline) with a specialized machine or worker at each station responsible for repeatedly performing simple tasks such as attaching the doors, welding joints, painting parts, and so on. These specialized machines or workers correspond to the specialized tasks in our pattern. Since the function of each station remains the same, we avoid the overhead incurred if we forced a station to switch roles as time progressed. Such a switch would require overhead to replace the existing role played by the station with a new one. In an automobile factory, this would mean either reprogramming robots or moving teams of specialized workers in and out of the stage.

Keeping individual stations performing the same role is more efficient. This greater efficiency is multiplied across the assembly line since each stage can specialize to a fixed task and a new car is produced at the end of the assembly line. The rate at which new cars are produced is dictated by the slowest stage in the pipeline. Why? If a slow stage feeds output into a faster stage, no matter how fast the second stage is it will always be limited by the rate at which it is fed work. The result of a situation like this is that the fast stage simply spends some time idle betwen producing a result and working on the next. Clearly ensuring that stages all take approximately equal times for each time step is critical in implementing a system that maximizes utilization and avoids wasteful idle time. Once this balancing has occurred, pipeline-based solutions provide a very efficient approach to solving problems that significantly increases throughput.

To understand the details of how a pipeline works, consider the pipeline illustrated in Figure 13.1. Assume that each step takes a fixed time of t seconds. At first, the pipeline is empty. When the first stage receives an input, it performs its small task and hands the result to the second stage after t seconds. The first stage receives a new input and performs its task on it at the same time that the second stage works on the output that it received from the first stage. Both complete after an additional t seconds. The result from the second is passed to the third stage, the result from the first to the second, and so on in a very regular pattern. Eventually the original input reaches the end of the last (nth) stage after $t * n$ seconds.

Now instead assume that we have a single complex station that could perform this entire task in some time $t * k$ where $k < n$ but $k >> 1$. While the pipeline increases overhead for a single item (due to the communication between stages), as the pipeline fills and the stages run concurrently, the performance benefit of pipelining becomes clear. For example, after $t * k$ seconds, the first result is produced. Immediately following it though, t seconds later

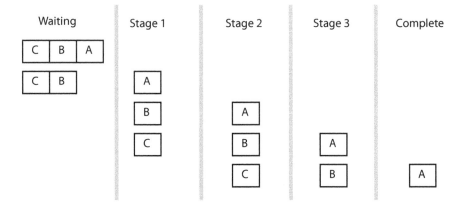

FIGURE 13.1: A simple illustration of a three stage pipeline showing a set of work units progressing through the pipeline.

is the result of the second input! Immediately following this, t seconds later is the next result, and so on. On the other hand, a single complex station would require $3(n * k)$ seconds to produce the results for three inputs. As the number of inputs grows larger and larger, the performance of the pipeline averages out to one complete result every t seconds — the performance is bounded not by the aggregate time for all stations in sequence, but for a single station in parallel with the others.

The benefits of pipelining can be profound for problems that naturally map onto an assembly-line style of solution in which a final result is reached by a sequence of smaller disjoint operations. This is why the approach is widespread in manufacturing. And for complex operations inside a microprocessor (such as floating point arithmetic) virtually every general purpose microprocessor available today uses pipelining.

13.1 Pipelining as a software design pattern

The pipeline design pattern is a direct translation of the assembly line concept into software. The key elements of the solution are:

- The original problem is decomposed into a number of specialized stages.

- The stages are connected by a flow of data. Usually this flow is a straight pipeline, but fan-in (multiple stages feeding into a single stage), and fan-out (one stage feeding multiple stages) can be used.

- Each data item input to the pipeline causes a data flow pattern that passes through all stages in a fixed order. Data flows in one end of the pipeline and flows out the other end, just as with a manufacturing assembly line.

- Other than data to control the flow of the computation, the stages are stateless, i.e., the state of the computation is contained in the data flowing between the stages.

In designing a pipeline, the stages ideally need to take an equal amount of time in order to balance the load among the processing elements of the parallel computer. Care must be taken to handle timing differences between stages and to prevent data from backing up between stages. Often, a handshake mechanism is needed so the input buffer of one stage doesn't overflow with data from a previous stage.

As mentioned earlier, pipelining is a common design pattern. Media processing in every form, from consumer electronics to scientific applications, make heavy use of this technique. For example, in processing a stream of video frames, a series of decompression and filtering stages are applied as video moves from a hard drive to a screen. The pipeline needs to support a throughput of a new frame every 16 milliseconds in order to provide a frame rate that is suitable for viewing as continuous video. As long as each stage in the pipeline takes less than 16 milliseconds (plus some additional leeway to cover the overhead of data movement) the overall target frame rate can be met. We gain concurrency up to the depth of the pipeline (i.e., the number of stages that execute concurrently) so the approach maps well onto GPUs and other parallel systems that are suited for streaming data through a sequence of pipelined computations.

13.2   Language support for pipelining

The pipeline pattern can be implemented in software in several ways. A programmer can use the SPMD pattern. The processes or threads supporting the SPMD pattern are specialized to their particular stages based on the process ID. The most challenging aspect of the problem (from a parallel programming point of view) is to set up the communication protocols to assure that communication buffers do not overflow. Overflows can occur when one stage is faster than that which it feeds into, leading to a buildup of pending work at the interface between the fast and slow stage. A simple handshaking protocol works, though it can lead to excessive overhead. Another approach is to use a queue data structure for the inputs into a stage. Then the handshaking can be relaxed and replaced with an event to stall a stage should an input buffer reach capacity. This approach means that if the stages are well balanced (i.e., take about the same amount of time) the overhead associated with the communication protocol should be minimal. Regardless of the approach taken, the pipeline pattern performs best when the stages are balanced and all take approximately the same amount of time.

While any language that supports basic communication and the SPMD pattern can be used to support the pipeline pattern, getting all the details implemented correctly and efficiently is non-trivial. Since this is such a common abstraction, higher-level abstractions to support the pipeline pattern make sense.

For signal processing and media applications, there is a long tradition in using visual programming languages to support pipeline algorithms (and their generalization to dataflow algorithms). Labview from National Instruments lets a programmer setup dataflow pipelines by drawing boxes on a screen and connecting them with lines.[1] Max/MSP from Cycling'74 is another system for pipeline and dataflow programming commonly used in electronic music and video applications.[2]

In both cases, concurrency is natural to exploit, but parallelism has not been a focus of these systems. This is changing, however, and in both cases the systems are evolving to increase opportunities to exploit concurrency and run well on multicore processors.

Functional programming languages are particularly adept at supporting the pipeline pattern. Pipelines ideally represent all state in terms of input and output flows of data. These "stateless" stages map perfectly onto the design goal of functional languages to eliminate side effects from functions. Many functional languages make queues easy to support in software which greatly simplifies the programming required to correctly connect stages in a pipeline. We will discuss the use of functional languages for the pipeline pattern in detail in the following sections when we consider the implementation of the pipeline pattern using Erlang.

13.3   Case study: Pipelining in Erlang

A language that provides language-level constructs that facilitate the construction of pipelined programs is Erlang. In Erlang, concurrency is realized by creating a set of processes that operate independently and communicate via message-passing channels. Each process is a single thread of execution in

---

[1]http://www.ni.com/labview/

[2]http://www.cycling74.com/products/max5/

Listing 13.1: Starting up the stages of a pipeline.

```
start() ->
    register(double,   spawn(fun double_loop/0)),
    register(add,      spawn(fun add_loop/0)),
    register(divide,   spawn(fun divide_loop/0)),
    register(complete, spawn(fun complete_loop/0)).
```

a non-shared memory context. Every process has a mailbox in which it can receive messages, and each process can send messages to other processes.

A pipeline is easily created in this model. You can imagine a pipeline stage as being nothing more than a process that performs its specific function on data elements that it receives in its mailbox, and the results that it produces are passed on to the next stage by simply sending a message to the corresponding process. How do we construct this sort of program in Erlang?

We will build a very basic pipeline where the following stages will be present.

- *Stage 1*: A number n is passed in, and the output is $2n$.
- *Stage 2*: A number n is passed in, and the output is $n + 1$.
- *Stage 3*: A number n is passed in, and the output is $n/3$.
- *Stage 4*: A number n is passed in, and the stage prints it.

The basic skeleton of a pipeline stage is composed of three parts: the code to receive incoming data from previous stages, the actual computational task to perform in the stage, and the code to pass the results of this computation to the next stage. There is also a small amount of code necessary to start up the processes that make up the pipeline and connect them together. We will start with this, as the pipeline is useless if it can't be constructed in the first place.

13.3.1   Pipeline construction

A pipelined program starts by first setting up the concurrent threads of execution that correspond to the pipeline stages. This process involves creating each thread and providing it with the contextual information necessary to know where data should be coming from and where it should send its results. In Erlang, this is quite easy. We will take advantage of the ability Erlang has for associating names with processes so that they can be looked up symbolically.

In Listing 13.1 we see the code to create each pipeline stage as an independent process. How do we read each of these lines? Working from the outside in, we see a call to the function **register** with two parameters — an atom that identifies a process, and the process ID that the atom refers to. We can think of this as providing a meaningful name that processes can use to look up the process IDs that they may wish to send messages to. To obtain this process ID, we have embedded the call to **spawn** in this second argument. **spawn** returns a process ID corresponding to the process that is created, and the argument to **spawn** is the function that forms the body of the process. For each of these stages we set the body function to *stagename*_loop, which is so named because the functions do nothing more than execute a loop of receiving messages that arrive in the mailbox of the processes and invoking the appropriate logic for the stage they represent.

13.3.2   Pipeline stage structure

A pipeline stage is responsible for three activities:

1. Receiving data from a previous stage.
2. Performing some work on that data.
3. Passing the results of the computation to the next stage.

The first step is generally performed by a **loop()** function within the module that represents a stage. In this loop, the process repeatedly waits for data from the previous stage, and upon receipt it invokes the computation it is responsible for with this received data. The result will then be passed to the next stage, **PidNext** as soon as the computation completes. We can implement this as:

```
loop() ->
  receive
    data -> PidNext ! computation(data),
            loop()
  end.
```

Let's look at this function to figure out what it is doing. The loop is composed of a single statement, **receive**. This statement blocks until a message arrives in the mailbox of this process. When messages are received that are of the form **data**, the computation is executed. The work routine **computation** is invoked with the data that was passed into the stage, and the computational work that the stage is responsible for executes. When the computation completes, the **computation** routine will return the data that is intended to be passed to the next pipeline stage. This result is immediately sent to the process identified as **PidNext**.

As soon as this message is sent, the stage immediately re-enters the **loop()** function waiting for more work. In Listing 13.2, we see the message handling loops for each of the stages. These correspond to the functions invoked when the stage processes were spawned by the **start** function. We also show an alternative implementation that can be used, where all stages use the same loop body and determine the role that they play based on the contents of the messages that they receive. The complication with this alternative implementation is that the messages must carry this additional baggage for allowing a stage to determine which role it should play for each message. On the other hand, this allows stages to be dynamically repurposed at runtime since they do not have a static role assigned to them for the lifetime of the process.

Finally, the computational functions that are invoked by the message handling loops are shown in Listing 13.3. We have placed print statements in the code for these functions so we can watch them execute in sequence. To start the pipeline, we simply need to send a properly formatted message to the first stage, `double`.

An interactive session showing compilation of the code (in a file called `pipeline.erl`), starting the pipeline, and injection of the value 4 to the first stage is shown below. Note that the "4" printed between the doubler and adder messages is due to the interpreter printing the result of the message passing operation, which is the value that was sent. The interactive Erlang shell and our pipeline stages are executing concurrently, so it isn't unexpected to see their output become interwoven like this.

```
Eshell V5.6.5   (abort with ^G)
1> c(pipeline).
{ok,pipeline}
2> pipeline:start().
true
3> double ! 4.
DOUBLER!
4
ADDER!
DIVIDER!
COMPLETE: 3.0
```

13.3.3   Discussion

The use of Erlang to construct pipelines is quite straightforward. We are simply responsible for defining a function for each stage that performs the computational work associated with the stage, and a loop handles receiving messages, performing the work, and passing the results on to the next stage. What are the language features that are assisting us here?

First, we have the transparent handling of the message mailboxes. Each Erlang process is provided by the runtime system with a buffer into which messages are received and stored. The programmer is not tasked *at all* with any details related to this activity — the messaging operation is transparent. The runtime system provides the machinery such that when a `receive` function is invoked, data is provided that may have arrived in the process mailbox. Furthermore, the semantics of the `receive` operation that dictate

Listing 13.2: Message handling loops for the stages of the pipeline.

```
%%
%% Message handling loops for each stage.
%%
double_loop() ->
  receive
     N -> add ! doubler(N), double_loop()
  end.

add_loop() ->
  receive
     N -> divide ! adder(N), add_loop()
  end.

divide_loop() ->
  receive
     N -> complete ! divider(N), divide_loop()
  end.

complete_loop() ->
  receive
     N -> completer(N), complete_loop()
  end.

%%
%% An alternative where one loop is present,
%% and the role a process plays in the pipeline
%% is determined by the contents of the message
%% that is received.
%%
loop() ->
  receive
     {double, N} -> add ! {add, doubler(N)}, loop();
     {add, N} -> divide ! {divide, adder(N)}, loop();
     {divide, N} ->
          complete ! {complete, divider(N)}, loop();
     {complete, N} -> completer(N), loop()
  end.
```

Listing 13.3: Computational bodies for the pipeline stages.

```
% stage 1 : double the input
doubler(N) -> io:format(''DOUBLER!~n''),
              N*2.

% stage 2 : add one to the input
adder(N) -> io:format(''ADDER!~n''),
            N+1.

% stage 3 : divide input by three
divider(N) -> io:format(''DIVIDER!~n''),
              N/3.

% stage 4 : consume the results
completer(N) -> io:format(''COMPLETE: ~p ~n'',[N]).
```

that the process should block until data is available is provided by the language. The programmer is not responsible for writing the code to poll and wait for data to arrive.

Similarly, the programmer is not responsible for implementing the message passing send operation either. The programmer simply defines the message it wishes to send, and hands this message to the runtime system using the exclamation point syntax to specify the message and the destination. The runtime ensures that the message is delivered, and the process itself is not concerned with any of the details as to how that happens.

13.4   Case study: Visual cortex

In some cases the pipeline pattern is found to naturally match the problem being solved without significant effort on the part of the software designer. In our earlier discussion of automobile assembly lines, a simulation of such a factory would be well suited to a pipelined structure. In this section, we discuss a similar simulation context where the structure of the process being simulated has a naturally pipelined structure well suited to this programming pattern. This simulation context is a model of the visual cortex in mammalian brains.

The processing of visual images by the mammalian brain is accomplished by spiking neurons grouped into several computational zones or layers. Each layer is responsible for processing information it receives and then feeding the

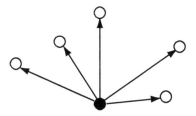

FIGURE 13.2: A directed graph of a spiking neuron (solid black node) connected to several post-synaptic neurons. The edges show the direction of the information flow.

transformed information on to the next neural layer through axons. From the perspective of computer science, neurons provide the computational units of the brain while axons provide the communication channels through which information travels.

Communication occurs when a neuron "fires" (spikes) and the resulting action potential travels over an axon until it reaches the synapse of another neuron, potentially causing the receiving (post-synaptic) neuron to fire in turn. This is shown schematically in Figure 13.2 where the spiking neuron communicates the spiking event with several other specific neurons.

Neurons are grouped into layers with each layer performing a set of computational tasks on its collective inputs. A simplified version of several layers connected in feed-forward fashion is shown in Figure 13.3. The neuron A in layer V1 receives its inputs from several other neurons in the layers below it. It then processes this information and passes the transformed information on to the next layer in the form of a temporal spike train. As can be clearly seen in the figure, the neural layers compose a pipeline, whereby an image presented to the retina is processed by the retina, then by the LGN, then by V1, and so on. In reality, there are also feed-back connections (not shown) from higher visual cortex layers to lower layers, but these are thought to be unimportant for visual processing times less than 25 milliseconds.

A simple feed-forward model of visual cortex could be easily parallelized by applying the pipeline pattern and assigning one or more neural layers to each core of a multi-core processor. Each layer contains a set of neurons n_i, a set of connections n_{ij} between neuron n_i in the current layer and neuron n_j in the previous layer, and a list l_i containing the index of the neurons that fire in a processing step. For example, consider Figure 13.4 where a one-dimensional

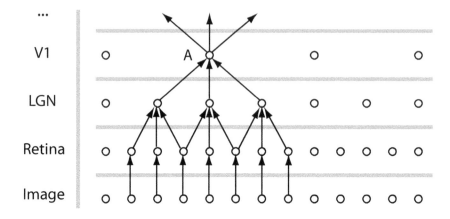

FIGURE 13.3:   A directed graph showing the layered structure of visual cortex connected in feed-forward fashion. Neuron A in layer V1 receives input from neurons in layers that are lower in the visual hierarchy.

slice of an image is presented to the retina. There are three on pixels (shown in black) in the image. In the first stage of the pipeline, the retina receives the activity list l_{img} from the retina, processes this information and passes a new activity list l_{ret} on to the LGN.

As described earlier in this chapter, the pipeline shown in Figure 13.4 won't improve performance if only a single static image is presented to the retina. However, our visual system is continually processing new information and the visual pipeline is a very efficient way to process large amounts of data in a massively parallel fashion. Recall that each neuron in visual cortex is a computational unit in itself, thus given roughly a billion neurons (10^9) all computing concurrently, the visual system has indeed solved the problem of parallel processing on a massive scale.

However, our simple pipelined, layered approach, with each processor core assigned one or more layers of neurons, is no slouch either. Contemporary processors operate in the GigaHz frequency range while neurons only compute in the KiloHz range (a factor of a million faster). So while the level of parallelism in our multi-core processors does not rival that of the brain, they can still process an impressive amount of information.

13.4.1   PetaVision code description

The pseudo-code necessary to implement the visual cortex model described above is now provided. For more detail, including a full-scale visual cortex

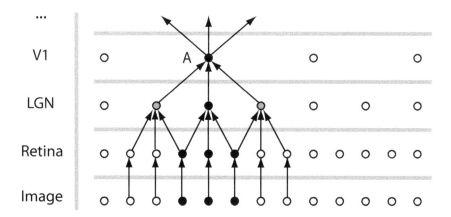

FIGURE 13.4: The flow of information as an image consisting of three on pixels is processed. The gray nodes have a lower probability of firing because they are affected less by the input from the image.

model, please visit the open source PetaVision project,[3] on which this code example is loosely based. This example describes a pipeline implementation on a single multi-core processor. In actuality, PetaVision is also apportioned using task parallelism with separate portions of a layer assigned to separate nodes (using MPI for communication between nodes).

The pipelined PetaVision model consists of a few important classes of objects. These classes are:

- A Layer class to manage neuron data and logic within a layer.
- A Connection class to maintain the connectivity graph between neural layers.
- A Column class to manage the layer pipeline.

The managing Column class and each layer run in a separate thread using shared memory for communication. The layers maintain their own internal state and only the spiking activity lists are shared between threads. Unfortunately, PetaVision is not implemented in a parallel language like Erlang, so it uses a low-level device called a mailbox for communication. The mailbox is an integer array with one element for each layer; it is assumed that a mailbox implementation can be written to atomically. The Column writes a command into a layer's mailbox but can only do so if the mailbox is available. A layer reads a command from its mailbox and changes the mailbox state to available once it has finished processing the command.

---

[3]http://sourceforge.net/projects/petavision/

Listing 13.4: Run method for Column class.

```
void run() {

  while a new input image exists {

    // signal layers to read and process inputs
    for all layers l {
      while (mailbox[l]) == NotAvailable) wait();
      mailbox[l] = RecvSynapticInput;
    }

    // signal layers to make firing decisions
    for all layers l {
      while (mailbox[l]) == NotAvailable) wait();
      mailbox[l] = UpdateAndFire;
    }
  }
}
```

Pseudo-code for the run method of the Column object is shown in Listing 13.4. It appears to be inefficient to wait for a layer's mailbox to become available before signaling the next layer. Indeed, will this pipeline really execute in parallel or will the threads only execute sequentially? It is left as an exercise for the student to consider these questions.

13.5   Exercises

1. Some problems have well known and static solutions. When these solutions are important to a large enough market, it makes sense to support them in hardware. A good example of this phenomena occurs in graphics. For a GPU of your choice, explore the use of the pipeline pattern inside the GPU. What specific architectural features were added to the chip to make these pipeline algorithms more effective?

2. Recast the matrix multiply problem from the exercises in Chapter 10 as a pipeline algorithm. How would the efficiency of this pipeline algorithm compare to the master-worker or data parallel versions? Are there situations on real parallel systems where the pipeline approach would be preferable?

3. Examine carefully the pseudo-code in Listing 13.4 used to signal a layer when to begin work. Will this pipeline execute in parallel or will the threads only execute sequentially? Why?

4. For the Listing 13.4 example, assume the workload for each stage is equally balanced, that it takes time $\tau_l = 0.5$ sec to process each command, and that $\tau_l \ll \tau_c$, where τ_c is the time the Column object takes to give the commands. Given 8 layers and 4 processor cores, how long does it take to process the first image? How long to process an image once the pipeline is full? How long to process 64 images?

Appendix A

OpenMP Quick Reference

OpenMP is an industry standard Application Programming Interface (API) for parallel programming. The language, created in the late 90s, has continuously evolved and is currently at the 3.0 level of revision. The most up to date information about OpenMP is available at `www.openmp.org`. OpenMP extends Fortran, C and C++ and was designed to make it straightforward for application programmers to create portable programs for shared memory computers (both SMP and NUMA). The basic idea is for programmers to incrementally add parallel constructs to their source code to evolve a serial program into a parallel program. Hence, the OpenMP constructs are designed to have a small (or even neutral) impact on the semantics of a program. OpenMP constructs are composed of a compiler directive and a block of code associated with that directive. In C and C++, the directive is in the form of a pragma:

```
#pragma omp directive-name [clause [ clause]]
```

The `directive-name` defines the basic construct while the optional clauses modify its behavior. The statement following the directive defines the block of code associated with directive. This block of code is called a "structured block." It can be almost any legal statement (or compound statement) in the host language. The major constraint is that the block must have one point of entry at the top and a single point of entry at the bottom. The only exception to this rule is an `exit()` statement.

The hope with OpenMP is that a single source code base can be used to support parallel and serial targets. If a programmer is careful, the directives in OpenMP can be semantically neutral. Hence, if a compiler that doesn't understand OpenMP is used, the pragma are ignored and a (hopefully) correct serial executable can be generated.

A.1   OpenMP fundamentals

To understand OpenMP, you need to appreciate three fundamental aspects of the language:

- Parallelism in OpenMP is explicit.

- OpenMP is a multithreading programming language.

- OpenMP at its lowest level is built on tasks.

We will consider each of these separately in the next few paragraphs.

First, parallelism in OpenMP is explicit. Programmers insert constructs into their source code which tell the compiler how to turn the program into a parallel program. While the compiler assists the programmer by managing the low-level details of how the parallelism is realized, it is up to the programmer to expose the concurrency in his or her program and express it for parallel execution.

Second, OpenMP is multithreaded programming language. OpenMP assumes that an OS will provide a set of threads that execute in a shared address space on behalf of the program. All of the issues common to multithreaded programming environments such as race conditions, livelock, and deadlock apply to OpenMP.

Finally, OpenMP is fundamentally built on top of tasks. Tasks are the basic unit of computation in OpenMP. A task defines the body of code to execute and the data environment required to support that execution. Tasks are defined by the programmer and scheduled for execution by the threads in the OpenMP program. Any time a thread is created, an implicit task is created as well. This implicit task is tied to the thread, i.e., the implicit task will only be executed by its associated thread. As we will see later, tasks can be explicitly created as well.

A.2   Creating threads and their implicit tasks

OpenMP is based on a fork-join model. A program begins as a sequential program. When code is encountered that would benefit from parallel execution, a team of threads are forked and the appropriate implicit tasks are created with the following directive:

```
#pragma omp parallel
```

Listing A.1: A basic parallel region.

```
#include <stdio.h>
#include <omp.h>

int main(int argc, char **argv) {
  #pragma omp parallel
  {
    printf("Hi!␣there.\n");
  }
}
```

Listing A.1 provides a simple example of a parallel region. This program creates a team of threads. The statement inside the parallel construct defines the implicit task which is executed by each thread. The result is that each thread independently prints a message to the standard output.

In Listing A.1, we did not specify the size of the team of threads. There are several ways to do this in OpenMP. The OpenMP environment creates a variable that keeps track of the number of threads to create should a parallel region be encountered. This variable is one member of a group of internal control variables, or *ICVs*. The value of the ICV is set by a runtime library routine

```
omp_set_num_threads()
```

read with the runtime library routine

```
omp_get_num_threads()
```

or assigned using the value of an environment variable

```
OMP_NUM_THREADS
```

This pattern is common for the other ICVs in OpenMP, i.e., an omp_get_, omp_set_ and environment variable.

In addition to the ICV, the number of threads to create for a particular parallel region can be specified with the num_threads clause. For example to request that the system create a team of 5 threads, you'd use the directive

```
#pragma omp parallel num_threads(5)
```

Listing A.2: Demonstrating shared and private variables based on scoping.

```c
#include <stdio.h>
#include <omp.h>

int main(int argc, char **argv) {
  int k = 10;

  #pragma omp parallel
  {
    int tnum = omp_get_thread_num();
    printf("Hi there (%d,%d)!\n", tnum ,tnum*k);
  }
}
```

A.3   OpenMP data environment

To understand the execution of tasks in OpenMP, you must understand the data environment seen by each task. OpenMP is a shared memory programming model. If data is defined outside a parallel region or if variables have global scope, they are shared between threads. If variables are defined inside an implicit task, they are private to that task.

For example, consider the program in Listing A.2. The variable k is defined outside the parallel region. It is a "shared" variable and is therefore visible to each of the implicit tasks tied to each thread. All tasks see the same value, 10. Inside the parallel region, an **int** variable called **tnum** is defined. Each implicit task tied to a thread has a private copy of **tnum**. The value of **tnum** in this case is assigned a value generated by the **omp_get_thread_num**() runtime library routine. This returns a number ranging from 0 to the number of threads minus one. This number can be used to uniquely identify each thread.

The program in Listing A.2 will print a string, its thread ID, and that ID times 10. The actual output statements are output in an arbitrary order based on how the threads are scheduled by the OS.

Private variables can be created with a clause as well. For example, in Listing A.2, if we wanted each thread to have its own copy of the variable k, we could add a private clause to the parallel directive

```c
#pragma omp parallel private(k)
```

Note that when you create a private variable, its value is undefined until it is explicitly assigned. If you want to create a private variable with a well defined value, you can use the **firstprivate** clause

Listing A.3: Demonstrating `threadprivate` variables.

```
#include <stdio.h>
#include <omp.h>

int counter = 0;
#pragma omp threadprivate(counter)

void hello(int tnum, int k) {
  counter++;
  printf("Hi there!(%d,%d).\n", tnum, tnum*k);
}

int main(int argc, char **argv) {
  int k = 10;

  #pragma omp parallel
  {
      int tnum = omp_get_thread_num();
      hello(tnum, k);
      printf("counter on thread %d = %d\n",
             tnum, counter);
  }
}
```

```
#pragma omp parallel firstprivate(k)
```

If this is used in Listing A.2, the result would be to give each implicit task its own value of the variable k, but now it would be assigned to the value held by the variable prior to the parallel construct (in this case, 10). We've discussed shared variable and private variables. There is an additional class of variables in OpenMP. These are `threadprivate` variables. These variables are private to a thread but shared between software modules within a thread. This is used with global scope variables in C or common blocks in Fortran. For example, in Listing A.3 we create a global scope variable called "counter" which allows each thread to keep track of how many times the function `hello()` is called by each thread. This example is trivial, but you can imagine a more complex example with many functions each incrementing the counter.

A private variable is also created when the reduction clause is used. A reduction is a common construct in parallel programming. A reduction occurs when a collection of computations are accumulated into a single value. To indicate a reduction in OpenMP, you use a reduction clause on an OpenMP directive

Listing A.4: Demonstrating a reduction.

```
#include <stdlib.h>
#include <omp.h>

int main(int argc, char ** argv) {
  int i, j, id, num;
  double mean = 0.0, *x;
  int N = 20000000;

  x = (double *) malloc(sizeof(double)*N);

  /* populate x on main thread ... */
  for (i=0; i<N; i++)
    x[i] = (double) rand() / (double) RAND_MAX;

  /* compute mean in parallel */
  #pragma omp parallel private(id, i, num) \
                       reduction(+:mean)
  {
    id = omp_get_thread_num();
    num = omp_get_num_threads();
    for (i=id; i<N; i+=num) {
      mean = mean + x[i] / (double) N;
    }
  }
}
```

```
reduction(op: variable)
```

where variable is a comma separated list of variables that appear in the associated structured block in an accumulation statement that uses the indicated
operator, op. For each variable in the list, a private copy is created. This private copy is initialized with the unity value appropriate to the operator (zero
for summation, one for multiplication, etc). Execution proceeds with each
thread accumulating values into its private copy of the reduction variable. At
the end of the structured block, the local values are accumulated into a single
value which is then combined with the global value. In Listing A.4 we show
an example of a reduction. The program will sum the values in an array to
compute the mean.

Notice the trick we used to distribute the loop iterations among the team
of threads. This is a common instance of the single-program, multiple-data
(SPMD) design pattern. We assigned the thread number to the private variable, id, and the total number of threads to the private variable num. The

loop limits were changed to run from the thread id to the number of items in the array, N, and we incremented the loop counter by the number of threads. The result is that the loop iterations are distributed round-robin between the threads, much as one would deal out a deck of cards.

The reduction clause

```
reduction(+:mean)
```

directs the compiler to create a private copy of the variable **mean** and assign it the value of zero. The summation inside the loop then occurs into this private value. At the end of the parallel region, the private values local to each thread are combined and added to the global value of the shared variable of the same name.

A.4 Synchronization and the OpenMP memory model

OpenMP is a shared memory programming language. The threads execute tasks within a shared address space. A memory model is called "sequentially consistent" if the threads assure that they all see the same memory updates in the same order "as if" the threads executed in a serial order. To maintain a sequentially consistent view of memory, however, adds a great deal of overhead. Hence OpenMP supports a relaxed memory model. The threads see updates to memory in a non-deterministic order depending on how the threads are scheduled for execution.

In most cases, a programmer doesn't need to think about the order of memory updates. But when managing the detailed sharing of data deep inside an algorithm, the fact that threads might have differing views of memory can create a problem. To resolve this problem, OpenMP supports a flush or a "memory fence." The basic flush directive is:

```
#pragma omp flush
```

This directive causes the calling thread to behave as if it wrote all shared variables to memory and then read them back. We say "as if" since the runtime system is free to work with the system to optimize this update to reduce memory traffic. A flush directive can take a list of variables as well so only the listed variables are updated to memory, but this is a dangerous form of the construct and we don't recommend using it.

Knowing when to use a flush can be challenging and to explain in detail how to safely use a flush goes way beyond the scope of this discussion. Fortunately, in the overwhelming majority of cases, a programmer can leave it to the runtime system to figure out when to use a flush. You do this by using the high-level synchronization constructs in OpenMP.

Synchronization is used to enforce constraints on the order of instructions among a collection of threads. The most basic synchronization construct in OpenMP is the barrier

```
#pragma omp barrier
```

A "barrier" indicates a point in a program at which all threads must arrive before any are allowed to proceed. As we will see, a barrier is implied at many points in OpenMP. For example, a barrier is implied at the end of a parallel region, i.e., all the threads wait at the barrier implied at the close of a parallel construct. When they all arrive, the master thread, that is the thread that originally encountered the parallel construct, continues while the other threads "join" (i.e., suspend and wait until the next parallel region).

The other synchronization constructs in OpenMP support mutual exclusion. When a team of threads encounter a mutual exclusion construct, the first thread there begins to execute the statements within the construct. The other threads wait until that thread is done. Then the next thread begins execution. This continues until one by one, each thread has executed the code within the construct.

The most basic mutual exclusion construct is a critical

```
#pragma omp critical
```

The structured block associated with the directive executes with mutual exclusion. For example, Listing A.5 shows the same problem addressed in Listing A.4, but instead of using a reduction, we manage the final accumulation with a critical section. We compute the local mean into the variable **pmean** and then accumulate the partial means into the global value inside a critical region. If we didn't use the critical region, the updates could conflict with each other and we would have a race condition. Also note on the parallel directive the use of the clause

```
shared(mean)
```

This tells the directive that the variable mean is to be shared between the threads. Technically, this clause is not needed for this program since "mean" is shared by default in this case. It is good practice, and can make the program easier to read, if shared variables are explicitly noted on the parallel directive.

OpenMP also defines an "atomic" directive:

```
#pragma omp atomic
```

The statement that follows the atomic directive is a simple operation to update the value of a variable. It is very similar to a critical construct, but was added to map more directly onto computer architectures that provide native support for an atomic update. The easiest way to understand atomic is by an example. Consider the code fragment

Listing A.5: Protecting updates to shared variables with a critical construct.

```c
#include <stdlib.h>
#include <omp.h>

int main(int argc, char ** argv) {
  int i, j, id, num;
  double pmean, mean = 0.0, *x;
  int N = 20000000;

  x = (double *) malloc(sizeof(double)*N);

  /* populate x on main thread ... */
  for (i=0; i<N; i++)
    x[i] = (double) rand() / (double) RAND_MAX;

  /* compute mean in parallel */
  #pragma omp parallel shared(mean) \
                       private(id, i, num, pmean)
  {
    id = omp_get_thread_num();
    num = omp_get_num_threads();
    pmean = 0.0;
    for (i=id; i<N; i+=num) {
      pmean = pmean + x[i] / (double) N;
    }
    #pragma omp critical
    mean = mean + pmean;
  }
}
```

```
#pragma omp atomic
x = x + rand();
```

The atomic construct only protects the read, update, and write of x. It is equivalent logically to the code

```
temp = rand();
#pragma omp critical
x = x + temp;
```

In other words, the execution of the function `rand()` occurs outside the atomic construct and hence it is not protected by a mutual exclusion relationship.

Finally, OpenMP includes low-level lock constructs. These give programmers the flexibility to build more complex synchronization logic into their program. The locks work just like the other mutual exclusion constructs, i.e., a thread acquires a lock and then proceeds to the following statements. If a thread encounters the lock and it is held by a different thread, it blocks until the lock is released.

Note that all the synchronization constructs in OpenMP imply a flush. So if you restrict yourself to the high-level synchronization constructs supplied by OpenMP, you can avoid mastering the intricacies of the OpenMP memory model and the use of the flush statement.

A.5   Work sharing

The fork-join model to create threads and implicit tasks is sufficient for a wide range of programming problems. The SPMD design pattern used with the parallel directive and the basic synchronization directives is sufficient for many problems. In this style of programming, however, the programmer must make many changes to the original sequential software. When you add the concept of an ID and a number of threads to a program, the program becomes specialized to a multithreaded execution. This violates one of the key goals of OpenMP.

Consequently, a number of constructs were included in OpenMP that spread work between a team of threads. The most common is the loop construct which in C has the form

```
#pragma omp for
```

which is followed by a basic for loop. The limits of the for-loop are simple integer expressions that use values that are consistent between all threads in a team. Consider the program in Listing A.6. This is our now familiar program

Listing A.6: Demonstrating a parallel for-loop.

```
#include <stdlib.h>
#include <omp.h>

int main(int argc, char **argv) {
  int i, j;
  double mean = 0.0, *x;
  int N = 20000000;

  x = (double *) malloc(sizeof(double)*N);

  /* populate x on main thread ... */
  for (i=0; i<N; i++)
    x[i] = (double) rand() / (double) RAND_MAX;

  /* compute mean in parallel */
  #pragma omp parallel shared(mean) private(i)
  {
    #pragma omp for reduction(+:mean)
    for (i=0; i<N; i++) {
      mean = mean + x[i] / (double) N;
    }
  }
}
```

to compute the mean of an array of numbers, but this time we use a loop construct.

The loop construct specifies that the iterations of the loop are to be distributed among the team of threads. The threads essentially share the work of computing the loop. There is an implied barrier at the end of the loop construct. The compiler has a great deal of freedom in choosing how to map loop iterations onto the threads. A programmer understanding his or her algorithm may want to influence how these iterations are mapped onto threads. This can be done with the schedule clause

```
#pragma omp for schedule(static, 10)
```

The static schedule takes as an argument a chunk size; in this example the chunk size is 10. The system divides the set of iterations into groups of size equal to the chunk size. These are then dealt out round robin among the team of threads. Another common schedule is the dynamic schedule

```
#pragma omp for schedule(dynamic, 10)
```

Once again the compiler blocks iterations into groups of size equal to the chunk size. In the case of a dynamic schedule, each thread starts with a block of iterations. When a thread finishes with its chunk, the runtime system schedules the next block of iterations for the thread. The runtime overhead is much greater for a dynamic schedule than for a static schedule. But when the work required for each thread is highly variable, the dynamic schedule can produce a much more balanced distribution of work among the threads. The combination of a parallel region followed by a work sharing construct is very common. Hence, OpenMP supports combining them in a single directive. We can replace the statements in Listing A.6 that define the parallel region and the parallel loop with the simpler code

```
/* compute mean in parallel */
#pragma omp parallel for private(i) reduction(+:mean)
for (i=0; i<N; i++) {
  mean = mean + x[i] / (double) N;
}
```

There are other worksharing constructs in OpenMP. They are used much less frequently than the loop construct, so we will only briefly mention them here.

A programmer can define distinct blocks of code that are mapped onto a thread's implicit tasks. Each block of code is called a "section." A set of these "section constructs" are combined into a larger group using a "sections" construct. For example, in Listing A.7 we show a program fragment where a team of threads are created to support a single sections construct. Inside this construct, we define three "section constructs" each of which executes a different function.

Listing A.7: Demonstrating sections.

```
#pragma omp parallel sections
{
  #pragma omp section
  do_x();
  #pragma omp section
  do_y();
  #pragma omp section
  do_z();
}
```

OpenMP also includes a "single construct." This defines a structured block that is executed by a single thread. There is no control over which thread in the team does the execution; just one of the threads executes the code in the structured block while the other threads wait. We show an example of this construct later in Listing A.8.

The work-sharing constructs imply a barrier at the end of the construct. This is the safest course of action and hence it is the default behavior. There are times, however, when the barriers in question would be extraneous. There are other times, when a programmer knows that given the data access patterns between work-sharing constructs that program will execute correctly without the barriers. In these cases, a `nowait` clause can be used

```
#pragma omp for nowait
```

The `nowait` clause should be used with great caution, however, as its improper use can introduce race conditions that could be very difficult to track down.

A.6   OpenMP runtime library and environment variables

OpenMP directives are based on pragmas in C. A compiler is free to ignore any pragma it doesn't understand. Hence, if a programmer only uses the directives in OpenMP, it is possible to write parallel OpenMP programs that behave correctly if compiled with an OpenMP compiler or with a compiler that doesn't understand OpenMP. In other words, the directives in OpenMP allow the programmer to support parallel and serial software from a single source code base.

Unfortunately, there are aspects of a parallel program that cannot be handled by a directive at compile time. We've encountered a few situations already where this is the case; namely when we need a thread ID and when we need to know the size of the team of threads.

```
omp_get_thread_num()
omp_get_num_threads()
```

In addition to these now familiar runtime library routines, there are collections of runtime library routines that manage the internal control variables (ICVs). As we explained earlier, these are variables internal to the OpenMP runtime that control how the program executes. Examples include the number of threads to use the next time a parallel construct is encountered. Others control if nested parallelism is allowed or if the number of threads can be dynamically varied between parallel regions. The pattern is the same in each case: there is one runtime library routine to "get" the value of the ICV, another to "set" the ICV, and an environment variable to set its default value. For example, to manage the ICV that controls if nested parallelism is enabled, we have:

```
omp_get_nested()
omp_set_neseted()
OMP_NESTED
```

There are other runtime library routines for locks, to modify the loop construct schedule at runtime, or learn about where you are within a hierarchy of nested parallel regions. An additional runtime library routine provides a portable interface to a clock.

```
double omp_get_wtime()
```

Each call to `omp_get_wtime()` returns a double precision value for the number of seconds that have elapsed from some fixed point in the past. You can bracket a block of code with calls to `omp_get_wtime()` to determine how much elapsed wall clock time has elapsed as the code in question ran.

A.7   Explicit tasks and OpenMP 3.0

OpenMP 3.0 was a major upgrade to the language. The most significant change was the addition of explicit tasks. Explicit tasks greatly expand the range of algorithms that can be handled by OpenMP. For example, below we show a code fragment with a while loop that traverses a linked list.

```
p = listhead; while (p) { process (p) p = next (p); }
```

Listing A.8: Demonstrating tasks.

```
#pragma omp parallel
{
  #pragma omp single private(p)
  {
    p = listhead;
    while (p) {
      #pragma omp task firstprivate(p)
                process (p)
      p = next (p);
    }
  }
}
```

This common control structure cannot be handled by any of the work-sharing constructs we have discussed so far. Explicit tasks allow one to parallelize this structure with straightforward directives. We show the code in Listing A.8.

As always, you must create the team of threads with a parallel construct. The single construct designates that a single thread executes the loop to traverse the list. Inside the loop, the task construct defines an explicit task. This occurs in two steps:

- *Packaging*: Each encountering thread packages a new instance of a task (code and data). In this example, the firstprivate clause indicates that each task is given a private copy of p which is initialized to the value in the global scope at the time the task is defined.

- *Execution*: Some thread in the team executes the task at some later time In essence, one thread (within the single construct) fills a queue of tasks. And the other threads in the team execute the tasks dynamically. When the thread creating the task queue is finished creating the tasks, it joins the pool of threads executing the tasks.

The task construct is very powerful. It supports while loops and other general iterative control structures plus graph algorithms, dynamic data structures, and a wide range of algorithms difficult to handle prior to OpenMP 3.0.

Appendix B

Erlang Quick Reference

Erlang is a mature declarative language that shares many features with functional and logic languages, such as Standard ML and Prolog. Erlang is unique in that concurrency constructs have long been part of the language, and the recent introduction of multicore processors into the market has fueled growth of interest in languages like Erlang. Erlang was created during the 1980s at Ericsson as part of an effort to design a language for distributed systems with a focus on robustness to failures and building parallel programs that communicate through explicit message passing.

In this appendix, we will focus primarily on the features of Erlang related to concurrency. Readers interested in the language are encouraged to refer to the Erlang Web site or books on the language to learn about the core language itself. An excellent reference to learn the language can be found in the recent book "Programming Erlang" [7]. Up-to-date versions of the compiler and language documentation can be found at the Erlang Web page, http://www.erlang.org/.

B.1 Language basics

Erlang programs are structured as modules which contain functions, some subset of which are exported and callable from outside the module. Functions that are part of a module and not exported are considered to be private within the module and inaccessible directly from the outside. If we wanted to build a module that is called "simple" that exports a single function "addOne" that takes a number as input and returns that number plus one, we would write the code shown in Listing B.1.

The first line establishes the name of the module. The second line exports a list of functions, where the square bracket notation represents a list. The contents of this list is the single entry addOne/1, meaning that the module

Listing B.1: A simple Erlang module.

```
-module(simple).
-export([addOne/1]).

addOne(N) -> N+1.
```

Listing B.2: Adding a second function to our module.

```
-module(simple).
-export([addOne/1, fact/1]).

addOne(N) -> N+1.

fact(1) -> 1;
fact(N) -> N * fact(N-1).
```

will export one function called **addOne** that takes one argument. Finally, we define the function **addOne**. Now, say we wanted to do something a bit more complex, like add a function to our module that computes the factorial of a number. We can modify the module as shown in Listing B.2.

What did we change? The first change was to add the second function to the list that we export. As you can see, the list on the export line now has two entries — one for **addOne** and one for **fact**. We also added the definition for the factorial function. Recall that the factorial of N is the product of all integers from 1 to N. We first define the base case, **fact(1)** representing the simplest instance of the function, since factorial of 1 is just 1. The semi-colon is used to say that we are not done defining the function just yet, and provide an additional definition that states when a number other than 1 is passed in, we should multiply that number by a recursive call to **fact** with the input minus 1. Note that we are not worrying about what would happen if a negative number were passed in at this point. We end the definition of factorial with a period stating that the function definition is complete.

Variables in Erlang must start with an upper-case letter, such as N above. This can be a source of confusion for programmers who step into Erlang from languages that do not impose this case requirement on variables. The reason for this is that Erlang also supports a type known as an *atom*, which starts with a lower-case letter. Variables are names for values, so in our **fact** function, if we pass 4 in, then N will correspond to the value 4. On the other hand, if we have an atom called **someatom** that occurs in our program, there is no value associated with it. The atom is simply a name. Atoms are commonly used to name things such as processes or annotations on data, while variables are used to name values. Beware of your cases when you start in Erlang and experience compilation errors.

Programs in Erlang (and other functional languages) very frequently use lists as a core data structure. A list is defined as a sequence of elements where we have a single element *head* and a list of elements that follow it called a *tail*. Lists are not to be confused with arrays, where we have random access to any element in an array in constant time. Lists only allow access to the head in constant time, and we must traverse the list to access elements that reside in the tail. Therefore if we create a list of integers called X,

```
X = [1,2,3,4,5,6].
```

then the code

```
[H|T] = X.
```

Would place the head of X, 1, in the variable called H, and the tail of X, [2,3,4,5,6], in the variable called T. The vertical bar (or "pipe") is used to distinguish the head from the tail of a list.

Finally, Erlang provides tuples for grouping data together. For example, if we wanted to write functions that work on two-dimensional vectors, it is good practice to represent them as a single entity containing two values instead of separating the components into distinct arguments to pass into a function. Say we wanted to add two vectors together, where each vector is composed of two values. We can do so by encapsulating the pairs of values in curly braces, { }.

```
addVec({XA,YA}, {XB,YB}) -> {XA+XB, YA+YB}.
```

Our function `addVec` then takes two arguments, each of which is a tuple that we are treating like a vector. We can mix types in tuples, including tuples themselves.

Given the basic syntax that we have just discussed, then we can build some interesting functions. For example, say we want to find the Nth element in a list, and return a special value that we will represent with the value `none` if the list doesn't contain an Nth element. We can build a module containing this function as shown in Listing B.3.

We start with the module name and export list, in this case seeing that our function `nth` takes two arguments. Our definition of the function is a bit more interesting that those we saw above. The first line, we see the case where the second argument is an empty list. Clearly if we are asking for the Nth element of an empty list we don't have one, so we return the atom `none`. The underscore for the first parameter means that we don't care what value it has, as the empty list in the second argument makes it unnecessary for us to look at it. It is syntactically valid if we had named it, but the compiler will emit a warning stating that we have an unused variable.

Listing B.3: Finding the *N*th element of a list.

```
-module(nth).
-export([nth/2]).

nth(_, [])      -> none;
nth(1, [H|_]) -> H;
nth(N, [_|T]) -> nth(N-1,T).
```

The second line defines the case where we are at the value we want — we want the first element of whatever list is passed in. Given that we have already dealt with the case where an empty list is passed in, we can assume that the list has a head which is the first element. So, we return it. Note that we aren't using the tail, so again, we use the underscore to tell the compiler that we don't care what the tail is to avoid a warning. Finally, the third line defines the case where a value of N is passed in that is greater than 1, so we recurse with $N - 1$ and the tail of the list and ignore the head since it is not the element we were looking for.

Frequently we want to do more with lists than just get out single elements. Consider the **merge** operation in the mergesort algorithm. This operation takes two sorted lists as input, and yields a single sorted list with their elements combined. It works its way down the lists looking at the first element of each, prepending the minimum of the two to the list formed by continuing this process on the remaining elements (with the head that was not the minimum reattached to its respective list). With a little help from the Erlang standard library, we can use the function **lists:split/2** to split a list in half and build a recursive mergesort that uses our **merge** function. Both of these are shown in Listing B.4.

In this example, we introduce a few new syntactic elements to our knowledge of the language that allow us to build even richer functions. We should also note that in this module, we only are exporting the **mergesort** function — the **merge** function is not exposed to the outside world.

First, consider the **merge** function. The first two definitions establish the case when one of the two lists is empty. In either case, we no longer have any merging to do, and return the list that did not match as empty. Of course, that doesn't mean that both lists can't be empty. What would happen in that case? The first definition would match the first list as empty, regardless of what the second list is. If the second list is empty, this would simply result in the empty list being returned — which is precisely what we want. The third and fourth definitions of **merge** get interesting. Examine the third definition carefully. Two interesting features are used here. First, we employ what is known as a guard to say that the third definition of **merge** corresponds to the case where we have two non-empty lists, and the head of the first list (**X**) is

Listing B.4: A simple mergesort module.

```erlang
-module(mergesort).
-export([mergesort/1]).

merge([],Y) -> Y;
merge(X,[]) -> X;
merge([X|TX],[Y|TY]) when X>Y ->
   [Y]++(merge([X|TX],TY));
merge([X|TX],[Y|TY]) ->
   [X]++(merge(TX,[Y|TY])).

mergesort([]) -> [];
mergesort([X|[]]) -> [X];
mergesort([X|T]) ->
   Middle = round(length([X|T])/2),
   {FRONT,BACK} = lists:split(Middle, [X|T]),
   merge(mergesort(FRONT),mergesort(BACK)).
```

greater than the head of the second list (Y). If the guard fails for a given pair of lists, then the fourth definition is used. The body of the third and fourth lines prepend the minimum head to the list created by recursively calling **merge** using the ++ operator. We can therefore read the third definition as stating that:

> When we have two nonempty lists where the head of the second list is smaller than the head of the first, we prepend a list containing the head of the second list (Y) to the result of recursively calling **merge** with the entire first list ([X|TX]) and the tail of the second (TX).

The **mergesort** function that uses **merge** is quite simple. First, we establish that sorting an empty list is just the empty list. Similarly, the list containing only one element is already sorted so we return it with no additional work to do. When we have a list with more than one element, we perform the recursive call to **mergesort** on the halves of the list that result from splitting the input at the midpoint. First, we determine the element where the split is to occur by dividing the length of the list by two and rounding the value to yield an integer. Both **round** and **length** are built in functions. We then use the **list:split/2** function, which takes the split point as the first argument and the list to split as the second to generate the two sublists that we wish to recursively process. Note that this function returns a tuple containing the halves of the list, which we call FRONT and BACK. Finally, we invoke **merge** on the results of recursively calling **mergesort** on the two sublists.

At this point, while we know only a sliver of the Erlang language, we can start to build interesting programs and explore the concurrency features of Erlang. For further information on the details of the full Erlang language, the reader should refer to [7].

B.2 Execution and memory model

Concurrency in Erlang is based on message passing between light-weight processes that do not share memory. As such, programmers in Erlang are not required to deal with many of the issues that arise in shared memory programming such as protecting critical sections and avoiding race conditions. The abstraction used by Erlang is often referred to as the Actor model (discussed in section 4.3.3), in which threads communicate asynchronously by deposing messages in mailboxes of other processes which can be checked for incoming messages at any time, regardless of when the message was delivered. This differs from synchronous message passing in which the receiver must be actively receiving a message at the same time as the sender transmits it, and either will block until the other arrives at the message passing transaction.

When an Erlang program starts, a single process is created and the program starts as a purely sequential process. To introduce concurrency, additional processes are created by using the **spawn** command. The **spawn** command takes a function as its argument, and returns a unique *process identifier*, or *PID*, as the result. For example, if we wish to spawn a separate thread to sort a list that we have called X and print the result to the screen, we could say:

```
X = [1,4,2,8,33,42,12,7].
Pid = spawn(fun() -> Sorted=mergesort:mergesort(X),
                     io:format("~p~n",[Sorted]) end).
```

In the spawn line, we use the **fun()** syntax to define a function of no arguments on the fly (sometimes called an anonymous function or a lambda), the body of which sorts the list and stores the result in a variable called **Sorted**, and then prints it using the Erlang function **io:format**. The comma separates the statements in the body of the anonymous function, and the **end** indicates the end of the function. These **spawn** then executes this function in a thread separate from the one which created it. The identifier of the spawned process is stored in **Pid** should we wish to use it.

When our process completes executing the function it goes away. If we wish the process to be persistent, then the function body must loop. This is typically achieved by recursion. We use this when implementing a concurrent pipeline in Chapter 13.

Given that the Erlang memory model is not a shared model, the means by which we construct processes that execute concurrently and interact with each other is via message passing.

B.3 Message passing syntax

Coordination between processes is achieved in Erlang programs through explicit message passing. Messages are passed by message sources sending data to a specific target, and receivers explicitly calling a `receive` function to acquire data that has been sent to them. Messaging in Erlang is asynchronous. This means that a message can be transmitted before the receiver has entered its receive call, and control continues past the send command on the sender side without waiting for the corresponding `receive` to occur. This is made possible by using the mailbox concept from the Actor model. Each process has a buffer known as a mailbox, and messages that are sent to each process are received and queued up in these buffers for later examination by the receiver. Control flow on the receiving side is not aware of incoming messages — the Erlang runtime system ensures that they will be received and stored for later reference regardless of what the receiving process is executing when the actual message arrives.

At the language level, Erlang provides very concise syntax for dealing with both the transmission and receipt of messages. To send a message `Msg` to a process with a PID `TargetID`, the sender uses the exclamation mark operator to state that the message is to be transmitted to the specified PID. In code, this is written as:

```
TargetID ! Msg.
```

On the receiving side, the `receive` function is executed. This function returns the contents of the first message in the mailbox. This message can contain any legal Erlang data structure. For example, let us consider the case where the sender sends messages that are tuples in which the first element is its own PID, and the second element is a value. It can send this as:

```
TargetID ! {self(), N}
```

And the receiver can act on it by executing:

```
receive
  {FromPID, N} -> doSomething(N)
end.
```

Listing B.5: The loop that our mergesort worker will execute.

```
loop() ->
  receive
    {FromPID, []} -> FromPID ! [],
                     loop();
    {FromPID, [H|T]} -> Sorted = mergesort([H|T]),
                        FromPID ! Sorted,
                        loop();
    {FromPID, _} -> FromPID ! notlist,
                    loop()
  end.
```

The receiving process executes the `receive` command, and the result is a tuple where the first element is the source of the message with the second argument being the actual data contained in the message. Erlang also provides mechanisms to name processes symbolically so that their PIDs can be discovered at runtime.

At this point, we can combine our knowledge of message passing with what we wrote earlier to build a simple process that listens for messages containing lists to sort, and returns them to the sender in a sorted form. Assuming we have our `mergesort` module from before, let's add to it one additional function. This function we will call `loop/0`, that takes no arguments and simply waits to receive a message. When a message is received, it acts on the message, passes the result back to the originator of the message, and recurses to wait for another request to arrive. This function is shown in Listing B.5.

Note that we are assuming `loop` resides in and is exported by the `mergesort` module. The body of the receive contains pattern matches for three different tuples that we can expect to receive. All of the tuples will be assumed to contain the PID of the originator of the message, `FromPID`. In the first match, we don't bother invoking `mergesort` with the empty list and simply send an empty list back to the sender. In the second match, we have a nonempty list, so we sort it and send the sorted list back. In the last case, we received a tuple that has something other than a list in the second position. We send an atom `notlist` to express this erroneous input back to the sender. In all three cases, we recursively call `loop` at the end to reenter the function and wait for more messages.

Listing B.6: Spawning a mergesort worker and interacting with it via messages.

```
SortPid = spawn(fun mergesort:loop/0).
SortPid ! {self(), [1,4,2,8,33,42,12,7]}.
receive
  N -> io:format("~p~n",[N])
end.
```

We can now create a process that will sort for us, and send it messages. The self() function can be used to include our PID in the tuple we send to the sorting process so that it can send the result back. This is shown in Listing B.6.

Appendix C

Cilk Quick Reference

Cilk is a multithreaded language derived from ANSI C [15, 14, 58]. The designers of Cilk state that it is also an *algorithmic* language. This is an important distinction to make relative to other multithreaded languages, as this means that the Cilk runtime guarantees both efficient and predictable performance. Cilk is based on both a language definition and a capable runtime system. The purpose of the runtime system is to remove the responsibility of thread scheduling, load balancing, and inter-thread communications from programmers, leaving them with the primary concern of identifying parallelism within their program.

The predictability of Cilk programs means that their analysis in a formal framework will yield predictions of performance that are close to those observed in practice. A programmer can analyze the performance properties of parallel programs based on two metrics: the *work* and *span* of the program. These metrics can be computed at runtime through automatic instrumentation inserted during compilation. This allows programmers to observe the actual parallel properties of their program under realistic workloads to understand how well the program parallelized and how it scales. Cilk achieves this through its work-stealing scheduler in order to dynamically balance work across processors as the program executes.

Programmers find that Cilk is very easy to pick up if they are already well versed in C, as the extensions Cilk makes to C are minimal and not disruptive to the base C language. Cilk was invented at the Massachusettes Institute of Technology and is the subject of a variety of published papers, articles, and student theses. As of the writing of this text (early 2008), Cilk had also been spun out of MIT into a commercial entity, Cilk Arts, which aims to produce a commercial grade version of Cilk known as Cilk++. In this appendix, we will provide a very brief overview of the original Cilk language. Those who find the base Cilk language to be of interest are encouraged to investigate Cilk++.

Cilk extends C in the following ways:

- Introduction of a small set of keywords to annotate program code with

Listing C.1: The basic Cilk example to compute Fibonacci numbers.

```
cilk int fib (int n) {
  int a, b;

  if (n < 2) {
    return n;
  } else {
    a = spawn fib (n-1);
    b = spawn fib (n-2);

    sync;
    return a+b;
  }
}
```

information related to parallelism.

- Addition of a small library of Cilk library routines.

- Usage of a *cactus stack* instead of the traditional C function stack for managing function invocation state.

To start the discussion of Cilk, refer to the example shown in Listing C.1. This is the example used by the Cilk developers to introduce the key features of Cilk, so we borrow it here for its simplicity. In section 12.2, we demonstrate a more sophisticated example based on the traditional mergesort algorithm for sorting lists of numbers. We also use Cilk in section 10.3 to construct a parallel program for computing images of the Mandelbrot set.

C.1   Cilk keywords

As we can see in Listing C.1, the Cilk version of the Fibonacci number generating function is identical to the equivalent recursive C function with a few keywords added that express where parallelism is created and synchronized. In fact, a single threaded instance of a Cilk program is known as the *serial elision* (or *C elision*) of the program, where the term *elide* describes the omission of the Cilk extensions that yields the pure C code.[1]

---

[1] *elision* [n.]: an act or instance of omitting something.

cilk

The first keyword of interest is `cilk`. This keyword is used to define Cilk procedures. A Cilk procedure is a routine that:

- May be spawned as a thread.

- May spawn additional threads.

- May synchronize upon completion.

- Is composed of C code.

A Cilk program may contain routines that are not annotated with this keyword that are legal to call from Cilk routines or non-Cilk routines. The constraint is that routines that are not annotated as Cilk procedures cannot participate directly in threading operations other than by being called from within a parent Cilk routine. A clear consequence of this is that the `main()` function of a Cilk program must itself be annotated as a Cilk procedure. Cilk also requires that `main()` have a return type of `int`.

spawn

Parallelism is achieved by using the **spawn** keyword to invoke Cilk procedures to execute in parallel with the parent. The **spawn** keyword is used to annotate what would otherwise be a simple function invocation, in which control is passed from the parent to a child function and the result of the child function would be stored in a variable specified by an assignment statement in the parent. The difference from a traditional function call is that the Cilk **spawn** annotation causes the child invocation to immediately return control to the parent while the child thread executes. The effect of this is that the parent is able to spawn other threads or perform its own computations while the child thread executes.

The fact that **spawn** immediately returns can lead to correctness problems if the parent thread uses the value that the child will assign to upon completion without first checking to see if the child has actually completed yet. Consider the code in Listing C.2.

In this example, the function `f(x)` spawns a child thread that executes `g(x+1)` and stores the result in the variable `a`. The parent, `f()`, computes the value of `b` based on the value of `a`. The problem is, the statement

```
b = a+1;
```

may execute before or after the child thread spawned to execute `g()` has completed. If `g()` has not completed yet, then `b` will take on the value 6, while if it has completed, then `b` will take on a value based on the result of `g()`. This sort of nondeterministic behavior is often undesirable, so Cilk provides the **sync** keyword to control it.

Listing C.2: Using spawned thread return values.

```
cilk int f (int x) {
    int a, b;
    a = 5;
    a = spawn g(x+1);
    b = a+1;
    return a+b;
}
```

Listing C.3: Using spawned thread return values with synchronization.

```
cilk int f (int x) {
    int a, b;
    a = 5;
    a = spawn g(x+1);
    sync;
    b = a+1;
    return a+b;
}
```

sync

The `sync` keyword is used to synchronize a parent thread with the completion of any child threads that it has spawned. As we saw earlier, the use by a parent of variables assigned to by a spawned child thread can lead to nondeterminism and the variety of correctness issues that results. The `sync` keyword is used within a Cilk routine to synchronize it with any child threads that it has spawned. We can limit the nondeterministic behavior of Listing C.2 by adding a single `sync` call before the fifth line of code, as shown in Listing C.3. This forces the parent to wait on the child thread before using the value that it returns.

A `sync` is a local barrier operation, which means that it applies only to the parent thread and the children that it has spawned. Cilk places an implicit `sync` before every `return` statement in addition to the explicit `sync` statements written by the programmer.

inlet and abort

In some cases it is useful to do complex things with values returned by spawned routines beyond simply storing them in a variable. Cilk provides `inlets` to implement these complex operations.

An inlet is a function that is internal to a Cilk procedure that takes as its

Listing C.4: A basic inlet to sum up an array.

```
cilk int summation(int *X, int n) {
  int sum = 0;
  inlet void add(int val) {
    sum = sum + val;
    return;
  }

  if (n==1)
    return X[0];
  else {
    add(spawn summation(X, n/2));
    add(spawn summation(&X[n/2], n-n/2));
    sync;
    return sum;
  }
}
```

argument the value returned from a spawned Cilk procedure. Consider the example in Listing C.4. In this case, we have a routine that computes the sum of an array of integers by recursively splitting the array in half, computing the sum for each half, and the overall sum by combining the results from each half. When each spawned child thread completes, its return value is added to the sum variable using the add() inlet. What makes inlets special is that within a single procedure invocation, all inlet invocations are atomic relative to each other. Given that inlet routines can contain quite complex code, the assurance of atomicity prevents correctness problems without requiring explicit guards to be put in place by the programmer.

Inlets also make possible a very interesting concept known as aborting. Say a procedure spawns off many threads to speculatively execute certain computations and learns that, after a subset of the computations complete, that the remaining threads should be terminated without completing. The abort primitive used within an inlet can be used to perform this termination operation. An abort within an inlet will terminate all existing children of the procedure containing the inlet. The programmer is responsible for maintaining state within the procedure (such as a flag) to prevent further spawns if the abort implies that they should not occur, and the programmer is responsible for cleaning up after the aborted threads.

C.2   Cilk model

Cilk is based on a concept known as *work stealing*. Given a set of processors, the Cilk runtime maintains a queue of threads assigned to each processor that have been created that are waiting to execute. As work units (threads) are completed on each processor, the next unit in the processor's queue is executed. When the queue that is associated with a processor is exhausted, the Cilk runtime will prevent the processor from sitting idle by stealing pending jobs from other processors that still have non-empty queues. The choice of processor to steal work from is random.

Threads that are created by **spawn** operations are pushed onto the queue of a processor, and execution of the threads corresponds to popping them out of the queue. Stealing occurs by removing a pending task from a nonempty queue on the side opposite that which the processor that owns the queue would itself ask for tasks. For example, if a processor retrieves tasks to execute from the front of its queue, a work stealing operation initiated by another processor would steal a task from the back of the queue. The reasoning behind this method of work stealing is based on the fact that tasks that exist in a work queue are likely to spawn tasks themselves. As such, a stolen task is likely to spawn children itself which would subsequently populate the task queue of the task thief. If the thief stole tasks at the other side of the source queue, the stolen task would be more likely to be related to already initiated spawned tasks already in the task queue.

C.2.1   Work and span metrics

The definition of Cilk as an algorithmic language that is built for analysis to understand performance and scalability of programs is based on the concepts of work and span metrics that can be computed for Cilk programs. Cilk programs are best thought of as directed acyclic graphs in which nodes represent threads and edges represent the spawning and termination of children. We show an example in Figure C.1 based on the Fibonacci computation in Listing C.1. Each wide group of three nodes corresponds to an invocation of `fib` in which more than one child invocation is spawned. The root group represents `fib(4)`, which has children corresponding to `fib(3)` and `fib(2)`, and so on. The groupings with single nodes represent the base cases `fib(1)` and `fib(0)`. Edges that point downward correspond to spawning of child threads, edges that point horizontally correspond to sequential execution within a thread, and edges that lead upwards represent threads completing and returning.

The first metric, **work**, represents the total amount of time for all threads in the program DAG. The work is referred to in the Cilk literature as T_1. This is equivalent to having only one processor on which all of the work is to be performed. In our figure, T_1 would be 17. The second metric, **span**, is

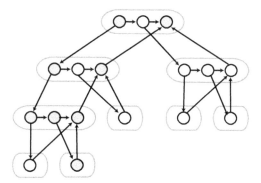

FIGURE C.1: The Cilk DAG corresponding to `fib(4)` based on an example from the Cilk tutorial by Charles Leiserson available on the Cilk Web page.

a measure of the longest dependency chain in the DAG assuming an infinite number of processors. The span is referred to as T_∞. The DAG in the figure indicates the nodes along this long chain as shaded, so T_∞ will be 8. Combining these measurements allows one to estimate the time required to execute a program on P processors as:

$$T_P \approx \frac{T_1}{P} + T_\infty$$

Cilk programs can be compiled with profiling instrumentation enabled allowing the collection of performance data to estimate the actual span and work observed at execution time. Compilation with the `-cilk-profile` and `-cilk-span` options will enable collection of this data for analysis after execution. Readers interested in exploring the performance analysis of Cilk programs are referred to the Cilk users manual and various papers about the development of the language.

C.2.2 Memory model

Cilk is based on a shared memory model. Cilk does not dictate any requirements for the target system with respect to the memory consistency model. This means that Cilk does not specify whether or not values are committed to main memory after an operation before the next operation executes. This is a decision deferred to the processor and underlying C compiler that Cilk itself uses for compilation. In the case that users write code where potential problems could occur, Cilk provides mechanisms to allow programmers to explicitly state where in the code memory operations *must* be committed before further operations are executed. Cilk does so via the `Cilk_fence()` primitive that is included as part of the Cilk standard library.

C.3   Cilk standard library

In addition to the annotations that Cilk provides as keyword extensions on the base C language, a set of standard library functions and macros are provided for programmers to use for finer control over synchronization and memory management. These are made available to programmers by including the header `cilk-lib.h`.

Memory allocation: `cilk_alloca`

The Cilk memory model is based on what is known as a *cactus stack* instead of the basic stack provided by C. This is necessary to provide for the set of call stacks that result from parallel function invocations. When a C program invokes a subroutine, it does so sequentially in a single stack. On the other hand, when Cilk executes, spawned threads may themselves make calls that result in a set of stacks associated with each parallel invocation. The cactus stack captures this behavior. The Cilk standard library provides a routine called `cilk_alloca` that allows for dynamic allocation of memory within the cactus stack. Memory allocated with `cilk_alloca` is automatically freed at the end of the function that invoked it.

Memory consistency: `Cilk_fence`

As mentioned earlier, it is occasionally necessary for programmers to explicitly state that memory operations must be committed at specific points in their code to ensure program correctness regardless of the memory consistency model enforced by the compiler or target processor architecture. The `Cilk_fence()` primitive is provided for this purpose. A consequence of this is that the `Cilk_fence()` primitive prevents compiler optimizations from moving code across the fence boundary during common code motion optimizations.

Locking

As discussed in the main text of this book, it is not unusual for programmers to come upon algorithmic instances where it is necessary to require a lock in order to protect some data visible between multiple threads in a shared memory context. When this cannot be easily achieved solely by `sync` operations to coordinate the completion of independent threads, Cilk provides a new data type for lock variables and a set of operations to act upon them. These variables can be declared with the type `Cilk_lockvar`.

These lock variables behave like typical locks. First, the programmer must initialize them with the `Cilk_lock_init()` function. Locks are not usable without first initializing them. Once a lock is initialized, the standard locking and unlocking operations are provided by the `Cilk_lock()` and `Cilk_unlock()` functions. The semantics of these operations are precisely what would be expected. If the lock is available when the `Cilk_lock()` function is invoked, the caller acquires the lock and is allowed to proceed. Otherwise, the lock function blocks until the thread that holds it releases it via the `Cilk_unlock()` function.

Synchronization testing: SYNCHED

When a thread has spawned a set of children, it can synchronize on their completion explicitly by blocking on a `sync` operation. In some cases though, this is not the ideal operation to use. Say a parent has a set of child processes that have been created, but can perform useful work even if only a subset of them have completed. It is wasteful from a performance perspective to block on a `sync` to wait for them to complete if useful work can be achieved even if they are not done yet. What the parent might like to do is to not block on their completion, but test if they have finished, and make a decision on further execution paths based on their completion status.

The `SYNCHED` special variable is available for this purpose. A parent can test this variable to see if children have completed. If they have not, instead of blocking as if a `sync` operation had been executed, the parent can choose to continue along an execution path in which useful work is achieved in the absence of the completion of the children. If a test of the `SYNCHED` variable succeeds, then an equivalent blocking `sync` operation would have immediately returned, so the body of the conditional testing `SYNCHED` would correspond with the function body that would follow the `sync` keyword.

The use of `SYNCHED` tests is a performance optimization used only when the programmer knows that useful work can be achieved if child threads are not ready yet at some point in the code. It is best to not use this special variable in practice until code has been created that works with a blocking `sync` call, and some excessive idle time is observed in the parent due to blocking that could be used by useful computations that are not reliant on the results of spawned children.

C.4   Further information

This appendix was written based on the version 5.4.6 of Cilk. Please check the Cilk documentation if you are using a more recent version in order to learn about new features and changes to those from the version discussed here.

- Cilk MIT Web Page: `http://supertech.lcs.mit.edu/cilk/`

- Cilk Arts Web Page: `http://www.cilk.com/`

References

[1] IBM Archives: 709 Data Processing System. http://www-03.ibm.com/ibm/history/exhibits/mainframe/mainframe_PP709.html.

[2] *CM Fortran Programming Guide*. Thinking Machines Corporation, January 1991.

[3] *MPI: A Message-Passing Interface Standard*. Message Passing Interface Forum, 1994.

[4] *MPI-2: Extensions to the Message-Passing Interface*. Message Passing Interface Forum, 1996.

[5] Gul A. Agha. Actors: A Model of Concurrent Computation in Distributed Systems. Technical Report 844, Massachusetts Institute of Technology, Artificial Intelligence Laboratory, 1985.

[6] Gregory R. Andrews and Fred B. Schneider. Concepts and Notations for Concurrent Programming. *Computing Surveys*, 15(1), March 1983.

[7] Joe Armstrong. *Programming Erlang: Software for a Concurrent World*. Pragmatic Bookshelf, 2007.

[8] John Backus. Can Programming be Liberated from the von Neumann Style? ACM Turing Award Lecture, 1977.

[9] John Backus. The history of FORTRAN I, II, and III. *SIGPLAN Not.*, 13(8):165–180, 1978.

[10] John G. P. Barnes. An overview of Ada. *Software–Practice and Experience*, 10:851–887, 1980.

[11] A. J. Bernstein. Analysis of Programs for Parallel Processing. *IEEE Transactions on Electronic Computers*, EC-15(5):757–763, 1966.

[12] Guy E. Blelloch. Scans as Primitive Parallel Operations. *IEEE Transactions on Computers*, 38(11):1526–1538, November 1989.

[13] Guy E. Blelloch. Prefix Sums and Their Applications. Technical Report CMU-CS-90-190, Carnegie Mellon University, 1990.

[14] Robert D. Blumofe. *Executing Multithreaded Programs Efficiently*. PhD thesis, Department of Electrical Engineering and Computer Science, Massachusetts Institute of Technology, September 1995.

[15] Robert D. Blumofe, Christopher F. Joerg, Bradley C. Kuszmaul, Charles E. Leiserson, Keith H. Randall, and Yuli Zhou. Cilk: An Efficient Multithreaded Runtime System. *Journal of Parallel and Distributed Computing*, pages 55–69, 1996.

[16] Peter A. Buhr, Michel Fortier, and Michael H. Coffin. Monitor Classification. *ACM Computing Surveys*, 27(1), March 1995.

[17] Nicholas Carriero and David Gelernter. How to Write Parallel Programs: A Guide to the Perplexed. *ACM Computing Surveys*, 21(3):323–357, September 1989.

[18] Bradford L. Chamberlain. *The Design and Implementation of a Region-Based Parallel Language*. PhD thesis, University of Washington, 2001.

[19] Bradford L. Chamberlain, David Callahan, and Hans P. Zima. Parallel Programmability and the Chapel Language. *International Journal of High Performance Computing Applications*, 21(3):291–312, August 2007.

[20] Bradford L. Chamberlain, Sung-Eun Choi, Steven J. Deitz, and Lawrence Snyder. The High-Level Parallel Language ZPL Improves Productivity and Performance. In *Proceedings of the IEEE International Workshop on Productivity and Performance in High-End Computing*, 2004.

[21] Bradford L. Chamberlain, Sung-Eun Choi, E. Christopher Lewis, Calvin Lin, Lawrence Snyder, and W. Derrick Weathersby. ZPL: A Machine Independent Programming Language for Parallel Computers. *IEEE Transactions on Software Engineering*, 26(3):197–211, March 2000.

[22] Keith Clark and Steve Gregory. PARLOG: Parallel Programming in Logic. *ACM Transactions on Programming Languages and Systems*, 8(1), 1986.

[23] Melvin E. Conway. Design of a Separable Transition-Diagram Compiler. *Communications of the ACM*, 6(7), 1963.

[24] George Coulouris, Jean Dollimore, and Tim Kindberg. *Distributed Systems: Concepts and Design*. Addison Wesley, 4th edition, 2005.

[25] David E. Culler and Jaswinder Pal Singh. *Parallel Computer Architecture: A Hardware/Software Approach*. Morgan Kaufmann, 1999.

[26] Steven J. Deitz. *High-Level Programming Language Abstractions for Advanced and Dynamic Parallel Computations*. PhD thesis, University of Washington, 2005.

[27] Edsger W. Dijkstra. Cooperating Sequential Processes. In F. Genuys, editor, *Programming Languages*. Academic Press, 1968.

[28] Edsger W. Dijkstra. Go To Statement Considered Harmful. *Communications of the ACM*, 11(3):147–148, March 1968.

[29] Edsger W. Dijkstra. Guarded Commands, Nondeterminacy and Formal Derivation of Programs. *Communications of the ACM*, 18(8), 1975.

[30] Edsger W. Dijkstra. Multiprogrammering en de x8. Circulated privately, n.d.

[31] Edsger W. Dijkstra. Over seinpalen. Circulated privately, n.d.

[32] R. D. Dowsing. *Introduction to Concurrency Using Occam*. Chapman and Hall, 1988.

[33] Kemal Ebcioğlu, Vijay Saraswat, and Vivek Sarkar. X10: An Experimental Language for High Productivity Programming of Scalable Systems. In *P-PHEC workshop, HPCA 2005*, 2005.

[34] Cândida Ferreira. Gene Expression Programming: A New Adaptive Algorithm for Solving Problems. *Complex Systems*, 13(2):87–129, 2001.

[35] M. Flynn. Some Computer Organizations and Their Effectiveness. *IEEE Transactions on Computers*, 21, 1972.

[36] Eric Freeman, Susanne Hupfer, and Ken Arnold. *JavaSpaces^{TM}: Principles, Patterns, and Practice*. Pearson Education, 1999.

[37] Erich Gamma, Richard Helm, Ralph Johnson, and John Vlissides. *Design Patterns: Elements of Reusable Object-Oriented Software*. Addison-Wesley, 1994.

[38] Jean-Luc Gaudiot, Tom DeBoni, John Feo, Wim Böhm, Walid Najjar, and Patrick Miller. The Sisal Project: Real World Functional Programming. Technical report, Lawrence Livermore National Laboratory, 1997.

[39] Al Geist, Adam Beguelin, Jack Dongarra, Weicheng Jiang, Robert Mancheck, and Vaidy Sunderam. *PVM: Parallel Virtual Machine*. MIT Press, 1994.

[40] S. Gill. Parallel Programming. *The Computer Journal*, 1(1):2–10, 1958.

[41] Brian Goetz. *Java Concurrency in Practice*. Addison Wesley, 2006.

[42] Thomas R. G. Green. Cognitive Approaches to Software Comprehension: Results, Gaps and Limitations. A talk at the workshop on Experimental Psychology in Software Comprehension Studies 97, University of Limerick, Ireland, 1997.

[43] Thomas R. G. Green and Alan Blackwell. Cognitive Dimensions of Information Artifacts: A Tutorial. BCS HCI Conference, 1998.

[44] Thomas R. G. Green and M. Petre. Usability Analysis of Visual Programming Environments: a 'cognitive dimensions' framework. *Journal of Visual Languages and Computing*, 7:131–174, 1996.

[45] Stuart Halloway. *Programming Clojure*. Pragmatic Bookshelf, 2009.

[46] Robert H. Halstead. Implementation of Multilisp: LISP on a multi-processor. In *Proceedings of the 1984 ACM Symposium on LISP and Functional Programming*, 1984.

[47] Robert H. Halstead. Multilisp: A Language for Concurrent Symbolic Computation. *ACM Transactions on Programming Languages and Systems*, 7(4), 1985.

[48] Per Brinch Hansen. Structured Multiprogramming. *Communications of the ACM*, 15(7):574–578, 1972.

[49] Per Brinch Hansen. The programming language Concurrent Pascal. *IEEE Transactions on Software Engineering*, 1(2):199–207, 1975.

[50] Tim Harris, Simon Marlow, Simon Peyton Jones, and Maurice Herlihy. Composable memory transactions. In *ACM Conference on Principles and Practice of Parallel Programming (PPoPP)*, 2005.

[51] Carl Hewitt, Peter Bishop, and Richard Steiger. A Universal Modular Actor Formalism for Artificial Intelligence. In *Proceedings of the International Joint Conferences on Artificial Intelligence*, 1973.

[52] Paul N. Hilfinger (ed.). Titanium Language Reference Manual, v2.20. Technical Report UCB/EECS-2005-15.1, University of California, Berkeley, August 2006.

[53] W. Daniel Hillis. *The Connection Machine*. MIT Press, 1985.

[54] C. A. R. Hoare. Monitors: An Operating System Structuring Concept. *Communications of the ACM*, 17(10):549–557, 1974.

[55] C. A. R. Hoare. Communicating sequential processes. *Communications of the ACM*, 21(8), 1978.

[56] M. Elizabeth C. Hull. Occam — A programming language for multi-processor systems. *Computer Languages*, 12(1):27–37, 1987.

[57] Jean D. Ichbiah, Bernd Krieg-Brueckner, Brian A. Wichmann, John G. P. Barnes, Olivier Roubine, and Jean-Claude Heliard. Rationale for the design of the Ada programming language. *ACM SIGPLAN Notices*, 14(6b):1–261, 1979.

[58] Christopher F. Joerg. *The Cilk System for Parallel Multithreaded Computing*. PhD thesis, Department of Electrical Engineering and Computer Science, Massachusetts Institute of Technology, January 1996.

[59] Simon Peyton Jones, Roman Leshchinskiy, Gabriele Keller, and Manual M. T. Chakravarty. Harnessing the Multicores: Nested Data Parallelism in Haskell. *Foundations of Software Technology and Theoretical Computer Science*, 2008.

[60] Ken Kennedy, Charles Koelbel, and Hans Zima. The Rise and Fall of High Performance Fortran: An Historical Object Lesson. In *Proceedings of the Third ACM SIGPLAN Conference on History of Programming Languages*, 2007.

[61] Donald E. Knuth. The Remaining Troublespots in ALGOL 60. *Communications of the ACM*, 10(10):611–617, 1967.

[62] Donald E. Knuth. *Selected Papers on Computer Languages*. CSLI, 2003.

[63] Donald E. Knuth and Luis Trabb Pardo. *The Early Development of Programming Languages*, volume 7 of *Encyclopedia of Computer Science and Technology*, pages 417–493. Marcel Dekker, 1977.

[64] Nancy P. Kronenberg, Henry M. Levy, and William D. Strecker. VAX-clusters: A Closely-Coupled Distributed System. *ACM Transactions on Computer Systems*, 4(2):130–146, 1986.

[65] Leslie Lamport. The Parallel Execution of DO Loops. *Commun. ACM*, 17(2):83–93, 1974.

[66] Leslie Lamport. How to Make a Multiprocessor Computer That Correctly Executes Multiprocess Programs. *IEEE Transactions on Computers*, c-28(9):690–691, 1979.

[67] D.H. Lawrie, T. Layman, D. Baer, and J. M. Randal. Glypnir — A Programming Language for Illiac IV. *Communications of the ACM*, 18(3), 1975.

[68] Berna L. Massingill, Timothy G. Mattson, and Beverly A. Sanders. SIMD: An Additional Pattern for PLPP (Pattern Language for Parallel Programming). In *Proceedings of the Pattern Languages of Programs Conference*, 2007.

[69] Timothy G. Mattson, Beverly A. Sanders, and Berna L. Massingill. *Patterns for Parallel Programming*. Software Patterns Series. Addison Wesley, 2005.

[70] John McCarthy. Recursive Functions of Symbolic Expressions. *Communications of the ACM*, 3(4):184–195, April 1960.

[71] John McCarthy, Paul W. Abrahams, Daniel J. Edwards, Timothy P. Hart, and Michael I. Levin. *LISP 1.5 Programmer's Manual*. MIT Press, 1965.

[72] James R. McGraw. The VAL Language: Description and Analysis. *ACM Transactions on Programming Languages and Systems*, 4(1):44–82, 1982.

[73] Robert E. Millstein. Control Structures in Illiac IV Fortran. *Communications of the ACM*, 16(10), 1973.

[74] Gordon E. Moore. Cramming More Components onto Integrated Circuits. *Electronics*, 38(8), 1965.

[75] Jarek Nieplocha, Vinod Tipparaju, Manojkumar Krishnan, and Dhabaleswar K. Panda. High Performance Remote Memory Access Communications: The ARMCI Approach. *International Journal of High Performance Computing and Applications*, 20(2):233–253, 2006.

[76] Rishiyur S. Nikhil. ID Language Reference Manual, Version 90.1. Computer structures group memo 284-2, Massachusetts Institute of Technology, 1991.

[77] Robert W. Numrich and John Reid. Co-Array Fortran for Parallel Programming. *ACM SIGPLAN Fortran Forum*, 17(2):1–31, 1998.

[78] Kristen Nygaard and Ole-Johan Dahl. The Development of the Simula Languages. *History of Programming Languages*, 1981.

[79] P. Nauer (ed.). Report on the Algorithmic Language ALGOL 60. *Communications of the ACM*, 6(1):1–17, 1963.

[80] David A. Padua and Michael J. Wolfe. Advanced Compiler Optimizations for Supercomputers. *Communications of the ACM*, 29(12):1184–1201, December 1986.

[81] C. V. Ramamoorthy and Mario J Gonzalez. Recognition and Representation of Parallel Processable Streams in Computer Programs. In *Proceedings of the ACM Annual Conference*, 1969.

[82] James Reinders. *Intel Threading Building Blocks: Outfitting C++ for Multi-core Processor Parallelism*. O'Reilly Media, Inc., 2007.

[83] Andreas Rossberg, Didier Le Botlan, Guido Tack, and Gert Smolka. Alice ML Through the Looking Glass. *Trends in Functional Programming*, 5, 2006.

[84] Michael L. Scott. *Programming Language Pragmatics*. Morgan Kaufmann, 2nd edition, 2005.

[85] Robert W. Sebesta. *Concepts of Programming Languages*. Addison Wesley, 8th edition, 2007.

[86] Nir Shavit and Dan Touitou. Software Transactional Memory. In *The Proceedings of the 14th ACM Symposium on Principles of Distributed Computing*, pages 204–213, 1995.

[87] Avi Silberschatz, Peter Baer Galvin, and Greg Gagne. *Operating System Concepts*. John Wiley and Sons, Inc., 8th edition, 2009.

[88] Sven-Bodo Scholz. Single Assignment C – Functional Programming Using Imperative Style. In *Proceedings of IFL '94*, 1994.

[89] Andrew S. Tanenbaum. A Tutorial on ALGOL 68. *Computing Surveys*, 8(2):155–190, 1976.

[90] UPC Consortium. UPC Language Specifications, v1.2. Technical Report LBNL-59208, Lawrence Berkeley National Lab, 2005.

[91] A. van Wijngaarden, B. J. Mailloux, J. E. L. Peck, C. H. A. Koster, M. Sintzoff, C. H. Lindsey, L. G. T. Meertens, and R. G. Fisker. *Revised Report on the Algorithmic Language ALGOL 68*. Springer-Verlag, 1973.

[92] Niklaus Wirth. Design and Implementation of Modula. *Software–Practice and Experience*, 7(1):67–84, 1977.

[93] Niklaus Wirth. Modula: A Language for Modular Multiprogramming. *Software–Practice and Experience*, 7(1):3–35, 1977.

[94] Stephen Wolfram. *A New Kind of Science*. Wolfram Media, 2002.

Index

abstract data types, 92
Actor model, 79, 160, 300
    Erlang, 160
Ada, 54, 128, 159
    entries, 128
    protected objects, 131
    rendezvous, 54, 128, 159
    select, 130
    tasks, 128
agenda parallelism, 200, 201, 247
ALGOL, 122, 163
    collateral clauses, 123, 163
AltiVec, 146
Amdahl's law, 26, 186
APL, 152
ARMCI, 71
array
    column-major ordering, 179
    data parallelism, 139
    expressions, 158
    notation, 152–158, 236–238
    row-major ordering, 179
    section, 158
    shift operator, 155
    slice, 158
    sparse, 152
    syntax, 158
    temporary, 154
assembly language, 87
asynchronous event processing, 65, 78
atomic, *see* atomicity
atomic regions, *see* critical sections
atomicity, 30, 188
    in sequential programs, 32
    issues with shared data, 32
    lock acquisition, 54
    of expressions, 90
    OpenMP atomic directive, 286
    shared counter example, 33
automatic parallelization, 145

batch processing systems, 111
Bernstein's conditions, 27
blocking, 189
busy wait, 55
bytecode, 105

C, 144
C++, 144
C#, 144
cache, 36, 112, 177–178
    coherence, 38, 181
    coherence protocol, 181–182
    issues in multiprocessors, 38
    latencies in a multilevel memory, 185
    memory hierarchies, 112–113
    miss overhead, 178
    multidimensional arrays, 179
cactus stack, 252, 312
call graph, 250
CDC 6600, 113
CDC 7600, 113
cellular automaton, 238–240
channels, 160, 165
Chapel, 3, 143
Cilk, 305–314
    abort, 309
    cactus stack, 252, 312
    directed acyclic graphs, 310
    fence, 311
    inlet, 226, 308
    Mandelbrot set computation, 226
    Master-worker, 226
    memory consistency, 311